C0-BKV-451

TESTABLE
FAITH

A REASONS TO BELIEVE ANTHOLOGY

Covina, CA

© 2022 by Reasons to Believe

All rights reserved. No part of this publication may be reproduced in any form without written permission from Reasons to Believe, 818 S. Oak Park Rd., Covina, CA 91724. reasons.org

Cover design: 789, Inc.
Interior layout: Christine Talley

Unless otherwise identified, all Scripture quotations taken from the Holy Bible, New International Version®, NIV®. Copyright © 1973, 1978, 1984, 2011 by Biblica, Inc.™ Used by permission of Zondervan. All rights reserved worldwide. zondervan.com. The "NIV" and "New International Version" are trademarks registered in the United States Patent and Trademark Office by Biblica, Inc.™

Names: Ross, Hugh, author. | Rana, Fazale, 1963-, author. | Samples, Kenneth R., author. | Zweerink, Jeffrey, author. | Haraksin, George II, author.
Title: Testable faith : a reasons to believe anthology / [by Hugh Ross; Fazale Rana; Kenneth Samples; Jeffrey Zweerink; and George Haraksin II.]
Description: Includes bibliographical references and index. | Covina, CA: RTB Press, 2022.
Identifiers: ISBN: 978-1-956112-02-3
Subjects: LCSH Intelligent design (Teleology). | Apologetics. | Religion and science. | BISAC RELIGION / Christian Theology / Apologetics | RELIGION / Religion & Science | SCIENCE / Life Sciences / Biochemistry | SCIENCE / Physics / Astrophysics | SCIENCE / Space Science / Cosmology
Classification: LCC BL240.2 R67 2022 | DDC 261.5/5--dc23

Printed in the United States of America

First edition

1 2 3 4 5 6 7 8 9 10 / 26 25 24 23 22

For more information about Reasons to Believe, contact (855) REASONS / (855) 732-7667 or visit reasons.org.

Contents

List of Figures and Tables

Figures

Tables

Acknowledgments

Hugh Ross, Fazale Rana, Kenneth Samples, and Jeff Zweerink hope you will see in each of the books from which this volume is derived their heart-felt expressions of gratitude to family members, friends, and colleagues. Many individuals assisted both personally and professionally in the writing, editing, designing, and publishing of those books. For this compilation of excerpts, their acknowledgments focus on the individuals who participated in bringing together what you now hold in your hands.

First, a word of appreciation goes to Baker Books, a division of Baker Publishing Group, for permission to include a chapter from *Why the Universe Is the Way It Is*. Baker has been a trusted partner over many years in bringing Reasons to Believe scholars' writing to light across North America and beyond. The team there has consistently been a pleasure to work with.

In the production of *Testable Faith*, our own teammates here at Reasons to Believe played major roles and deserve great thanks. Thank you to Bryan Rohrenbacher, Brian Bowman, Elissa Fernandez, and their respective teams for helping bring this book to life.

Deep gratitude goes to our editorial team—Sandra Dimas, Joe Aguirre, Maureen Moser, Brett Tarbell, Helena Heredia, and Jocelyn King—who polished and refined the draft. The creative skills of the team from 789, Inc., shine from the cover design and those of Christine Talley from the book's interior. We appreciate your talents!

Finally, we would like to acknowledge all our other colleagues at Reasons to Believe. Their support, encouragement, and effort go far beyond the call of duty. We're blessed to work with men and women who share our passion for providing new reasons to believe in our Creator and Savior.

Introduction

by George Haraksin II

Imagine you're at a dinner party with people you've never met. As the night goes on, you get to know the guests, listen to their experiences, and engage in conversations on a wide variety of subjects. This book is a bit like that party. The guests include a biochemist, two astrophysicists, and a philosopher-theologian (staff scholars at Reasons to Believe [RTB]), and the topics span across multiple disciplines. Each chapter, pulled from previously published works, explores pressing questions that put the Christian faith to the test.

But is faith *testable?* Some might think that faith is believing what you know isn't true, but this is not the biblical notion of faith and truth. In the Gospels, Jesus often speaks of the spiritual realm, a reality beyond the physical or natural world, yet he also maintains that empirical evidence is available for many of his affirmations.[1] In Matthew's Gospel, while jailed John the Baptist sends some of his disciples to raise questions about Jesus's legitimacy. When Jesus responds to the inuiry, he points to observable, testable evidence saying, "Go back and report to John what you <u>hear</u> and <u>see</u>: The blind receive sight, the lame walk, those who have leprosy are cleansed, the deaf hear, the dead are raised, and the good news is proclaimed to the poor" (Matthew 11:4–6).[2] Jesus assumes that a person's beliefs "ought to fit the facts," as philosopher Douglas Groothuis puts it, and he is willing to appeal to testable evidence to support his claims.[3]

If we are called to test all things, then we can expect that, when interpreted rightly, God's two books—the book of nature (creation) and the book of Scripture (the Bible)—would be in harmony. This "two books" approach attempts to demonstrate that the biblical creation model is more plausible and better fits the facts—facts derived from both Scripture and nature—than other models. The model is testable, can be assessed by empirical evidence, and can be improved as we encounter apparent conflicts or tensions that arise between scientific and biblical interpretations of the data.

Today, such a model is needed more than ever. While the perceived rift between science and faith certainly isn't new, the percentage of those who believe that science and religion are in conflict continues to increase. According to Pew Research, "73% of adults who seldom or never attend religious services say science and religion are often in conflict, while half of adults who attend religious services at least weekly say the same."[4]

RTB exists to help change this narrative.

Our *passion* is to use science apologetics in the service of evangelism. In the words of late Christian philosopher Dallas Willard, "we do [apologetics] as disciples of Jesus; in the manner in which [Jesus] would do it. We do it to help people . . . Apologetics is a helping ministry. . ."[5]

This help comes through demonstrating that God's two books are not in unrelenting conflict like a pair of rabbling foes but can live in a harmonious relationship that boldly faces (and withstands) tensions as they inevitably arise. We aim to show that science and the Christian faith grow into deeper unity as we discover truth and attain real knowledge, which includes knowledge of God and the gospel of his kingdom (Colossians 1:10).

RTB's passion is supported by the substance of our *positions*, presented through our testable creation model. This model, which we continue to refine with helpful evaluation from colleagues and critics, displays RTB's understanding and interpretations of the best available scientific evidence regarding the origin of the universe, the origin of life, human origins, and the apparent design and complexity of life. Our testable creation model contains various predictions on what science is likely to discover in the future. The model is built on sound science and offers testable truth and a testable faith, as displayed in these pages.

In this book, philosopher-theologian Kenneth Samples starts off the conversation by tackling two pressing questions that Christians and non-Christians ponder: isn't faith incompatible with reason; and, if God created everything, then who created God? Astrophysicist Hugh Ross then shifts the discussion to the cosmos to address the next questions an inquisitive mind might ponder: how did the universe begin and why is it so vast? In chapter 5, astrophysicist Jeff Zweerink dives deeper into the topic of the cosmos by exploring its design. Then, in chapter 6, he turns his focus to extraterrestrials—do they exist and would their existence disprove Christianity? Biochemist Fazale Rana wraps up the discussion by sharing why he believes God exists. He concludes by making a case for the image of God.

Throughout these pages, and in the work of RTB, we strive to display, maintain, and cultivate a *posture* of gentleness and respect toward others that accords with the apostle Peter's encouragement in his first biblical letter (1 Peter 3:15). In maintaining such a posture, we aim to help people come to love God with all their mind and to gain knowledge using sound science that leads us to testable truth and testable faith.

Our goal is that, by the end of this book, you will have gained a deeper understanding of the "two books" approach to understanding the Bible and creation. As RTB founder Hugh Ross likes to say, "God gave us the book of nature and the book of Scripture to show us who he is . . . to show us his role in our world and in our lives . . . to reveal his desire for a personal relationship with each of us."

We invite you to take a seat at the table and explore how strong faith and sound science can and do walk hand in hand.

Chapter 1

Isn't Faith Incompatible with Reason?
by Kenneth Richard Samples

> Belief in God is not irrational but possesses its own distinct
> and robust rationality. It represents a superb way of making
> sense of things.
>
> —Alister McGrath, "Isn't Science More
> Rational than Faith?"

Christians are often charged with the claim that faith is unreasonable and that those who believe in it are driven by an emotional need for security. For example, the late Christopher Hitchens declared, "Faith is the surrender of the mind, it's the surrender of reason, it's the surrender of the only thing that makes us different from other animals."[1] Richard Dawkins calls faith "the great cop-out."[2]

It is easy to think that this is a recent accusation from twenty-first-century atheists, but this criticism goes all the way back to the ancient world. Second-century Greek philosopher Celsus accused Christians of being—among other things—ignorant, unintelligent, uninstructed, and foolish people.[3] In Celsus's mind, only the unintelligent people became Christians; smart, thoughtful people were not taken in by it. Most Christians in the ancient world were not formally educated, but the same was true of most people of the time. From the very beginning, well-educated people have believed in Christ. (The apostle Paul especially comes to mind.) Moreover, Scripture and Christian philosophy consistently relate faith to reason in powerful ways. The coherent explanatory power and scope of the Christian world-and-life view has shaped the entire intellectual culture of Western civilization.

Nevertheless, many people perceive faith and reason as being at odds with one another. For example, some differentiate faith from reason by asserting that *faith* merely involves *hoping* something is true, whereas *reason* involves *affirming* something to be true based upon *justifying evidence*. According to this model, faith is equivalent to mere wishful thinking and is, thus, incompatible with reason.

People often use the term *blind faith* to describe religious belief. As Dawkins puts it, "Faith is belief in spite of, even perhaps because of, the lack of evidence."[4] Blind faith is confidence without any good reason for such self-assurance. From this perspective, faith is sightless and without justification and therefore, again, is incompatible with reason. Unfortunately, there are times when Christians appear, whether in word or deed, to engage in or encourage blind faith. However, to assume that this behavior—real or misinterpreted—represents Christianity's true relationship with reason is a grave mistake.

Historic Christianity Relates Faith to Reason

Let's now look at nine ways that historic Christianity connects faith with reason.

Trust in a Credible Source

Christianity's general view of faith and reason is very different from the popular stereotypical definition. Reason is respected, even prized, within the Christian tradition. A powerful theological-philosophical consensus within the history of the faith has argued that the Christian religion involves knowledge, is compatible with reason, and is even the product of reason. This agreement has often been expressed in the common statement, "faith seeking understanding." Its most articulate and persuasive spokespersons through the centuries have been distinguished thinkers such as Augustine, Anselm, and Thomas Aquinas.[5]

Biblical faith isn't synonymous with wishful thinking. The root words for *faith* in both the Old Testament (Hebrew) and New Testament (Greek) mean "trust" (see table 1.1). These biblical terms convey a confident reliance on someone or something. But that trust must be placed in a *credible* (reasonable and/or reliable) source or object. One doesn't place their trust in someone or something they know nothing about. In other words, biblical faith is never blind. Rather, for both traditional Jews and Christians, the believer must know that the object of their trust is in fact *trustworthy*. Faith in the biblical sense is confident *trust* in a *credible* source; that source can be God, Christ, the truth, a parent or teacher, etc. Thus, scripturally speaking, rather than being in opposition to reason, faith's very definition includes a necessary rational component.

To put it philosophically, like the traditional definition of knowledge, faith

Table 1.1 Basic Biblical Words for "Faith"[6]
Hebrew: *'emun*, means "faithfulness or trustworthiness" Greek: the verb, *pisteúō*, means "to believe"; the noun, *pístis*, means "faith or trust"

involves justified or warranted belief. It involves a reasonable step of trust or confidence in what a person has good reason to believe is indeed true. Similarly, philosophers commonly define knowledge as "justified, true belief." In other words, a person possesses knowledge when they believe something that is true and is accompanied by proper justification (facts, evidence, reasons).[7] According to these definitions, faith includes a necessary rational component and knowledge includes a necessary faith component. A person must *believe* something to *know* anything.

In Scripture, faith involves knowledge. For example, saving *faith* depends on knowing certain historical facts about Jesus's life, death, and resurrection. Christians have faith in Jesus Christ by believing he is the Messiah (God's specially anointed servant), the incarnate Son of God, and the crucified and risen Savior of the world.

New Testament writers insist on the credibility of their accounts, correcting false beliefs about Jesus on the basis of that credibility. The apostle Peter explains, "For we did not follow cleverly devised stories when we told you about the coming of our Lord Jesus Christ in power, but we were eyewitnesses of his majesty" (2 Peter 1:16). The apostle Paul, arguably the greatest apologist for Christ, declares, "We demolish arguments and every pretension that sets itself up against the knowledge of God, and we take captive every thought to make it obedient to Christ" (2 Corinthians 10:5). Moreover, the New Testament authors also demonstrate that justification for believing the truth of Jesus's identity comes through such things as Christ's fulfillment of prophecy, his miracles, his moral perfection, and his bodily resurrection from the dead. For example, Matthew's Gospel quotes Old Testament prophets to establish Jesus's identity as the expected Messiah (Matthew 1:22–23; 2:6, 15, 18; 3:3; 21:4–5). The Christian faith is directly connected to the rational *knowing* process.

Faith Seeking Understanding

In historic Christianity, reason and faith function in a complementary fashion. Apart from God's special grace, reason in and of itself cannot cause faith[8]— nevertheless, reason normally plays some role in a person's coming to faith and afterward supports faith in innumerable ways. As we've established, even the very faith that results in salvation involves knowledge and discursive reasoning (logically working toward a conclusion). Saving faith includes knowledge of the gospel, assent to its truth, and confident reliance on the Lord and Savior Jesus Christ. It incorporates a human's full faculties in the innermost part of the person—mind (knowledge), will (assent), and heart (trust). Christianity combines faith and reason in a harmonious way that enhances the qualities of each.

Both Scripture and long-standing tradition teach Christians to seek rational understanding of their faith. While faith itself is a gift of divine grace, reason can evaluate, confirm, and buttress it. A well-reasoned faith stands a better chance of enduring disturbances and overcoming challenges. For example, in his letter to the Colossians, Paul warns against falling for "hollow and deceptive philosophy" (Colossians 2:8). Understanding the rational foundations of one's faith is an excellent antidote to false teaching.

Reason is also vital to apologetics and evangelism. Peter tells Christians to "always be prepared to give an answer to everyone who asks you to give the reason for the hope that you have" (1 Peter 3:15). (This verse contains the mandate for the Christian apologetics enterprise.) Paul used reason extensively in his efforts to share the gospel with non-Christians. The book of Acts describes him engaging in debate with intellectuals in Athens (Acts 17:16–34) and reasoning from the Scriptures in local synagogues (Acts 17:2).

In the centuries following Christ's resurrection, many world-class thinkers have professed faith in Jesus, among them Augustine, Anselm, Thomas Aquinas, and Søren Kierkegaard. These men and others reinforced the complementary relationship between faith and reason through their philosophical and theological writings (see table 1.2).

Reason and the Mysteries of Faith

As finite creatures with real limitations and boundaries, humans can't fully fathom the divine mysteries of Christianity such as the Trinity and the incarnation. Nevertheless, we can define these doctrines in ways that promote comprehension while avoiding actual logical contradictions (absurdities). For example, the oneness (essence) of the triune God is in a different logical respect from his

Table 1.2 Historic Christian View of Faith and Reason

The Christian church has a history of diverse views regarding the proper relationship between faith and reason. There is, however, much common ground between these views. The expression "faith seeking understanding" best captures the spirit of the consensus of Christian philosophy.[11] Here's a summary of some of the great Christian thinkers' perspectives on this topic:

- Augustine (354–430): "Believe in order that you may understand" (*Crede ut intelligas*). For Augustine, faith and reason have an interdependent relationship and both are uniquely enabled by divine grace. Augustine's views have had a monumental influence over the philosophies of his successors, including those included in this list.
- Anselm (1033–1109): "I believe in order that I might understand" (*Crede ut intelligam*). Anselm laid emphasis upon faith being prior to reason and understanding.
- Thomas Aquinas (1225–1274): "I understand and I believe" (*Intelligo et credo*). Aquinas believed that some truths are discovered through both faith and reason, whereas other truths are known exclusively through faith (special revelation). Nevertheless, human reason is finite and negatively impacted by sin, so grace buttresses both.
- Søren Kierkegaard (1813–1855): "I believe because it is absurd" (*Credo quia absurdum est*). Kierkegaard emphasized the importance of faith and that the gospel message (God taking on a human nature to atone for human sin) is an affront to human autonomous reason (i.e., seemingly unreasonable to non-Christians), but his view need not imply a rejection of reason or that Christianity is actually absurd.[12]

threeness (personhood). God can be defined as one essential *What* and three personal *Whos*. This definition avoids violating the law of noncontradiction.[9]

Christian truths aren't irrational or absurd, though they often transcend finite human comprehension. In other words, faith doesn't damage or violate reason itself. When skeptics have challenged the logical coherence of biblical teachings, Christian thinkers through the centuries have offered viable models for showing these teachings to be mysterious, but not actually incoherent.[10]

God's Rational Mind and Reason
The Christian worldview offers a plausible explanation for affirming an objective source for knowledge, reason, and rationality. That explanation is found in a personal and rational God. Christian theism affirms that an infinitely wise and all-knowing God created the universe to reflect a coherent order of laws and logic (Greek: *nómos*, "law"; *lógos*, "logic"). He also created humans in his image (Latin: *imago Dei*) and endowed them with rational capacities to discover that coherent order (Genesis 1:26–28). In effect, God networked the comprehensible cosmos and rationally capable humans together with himself to allow for a congruence of intelligibility. Thus, in Christianity, the rationality of the universe has a reliable metaphysical foundation in the infinite and eternal mind of the Creator God. Likewise, human finite minds are the rational product of God's infinite mind. Thus, one can say that Christian truth claims themselves are the product of divine reason.

The Bible and Intellectual Virtue
Christians are often portrayed in media and entertainment as anti-intellectual and, unfortunately, there are some quarters of the faith that feed this stereotype. Anti-intellectualism is not what the Bible teaches. Scripture encourages the attainment of knowledge, wisdom, and understanding (Job 28:28; Psalm 111:10; Proverbs 1:7). The Old Testament features an entire genre centered on wisdom. Scripture also promotes such intellectual virtues as source-checking, discernment, testing, reflection, and intellectual renewal (Acts 17:11; 1 Corinthians 14:29; Romans 12:2; Colossians 2:8; 1 Thessalonians 5:21). The apostle Paul teaches that the Christian life should be marked by what he calls the "renewing of your mind" (Romans 12:1–2). Jesus sums up the gist of Mosaic law as "Love the Lord your God with all your heart and with all your soul and with all your mind" (Matthew 22:36–40). This involves striving to use one's cognitive faculties to their fullest extent in one's devotion to God.[13]

Christianity and the Rational Investigation of Truth
Christianity invites people to investigate and test its rational truth claims. It is a religious tradition that can be either potentially verified or falsified. The New Testament provides various examples of promoting and appealing to the rational investigation of Christianity's truth claims. Consider four examples:

1. Jesus Christ himself invited people to consider and assess evidence concerning his own identity as the divine Messiah (Matthew 16:13–17; John 14:11; Acts 1:2–3).
2. The apostle Paul appeals to evidence for the resurrection of Jesus as establishing the truth of the faith (1 Corinthians 15:3–8).
3. The apostle John says that he was a witness to the life and ministry of Jesus and testifies to the truth of the faith (1 John 1:1–4).
4. Luke asserts that he has investigated the claims of Christianity and sets forth an orderly report (Luke 1:1–4).

Christianity and Rational Disciplines
In the last 2,000 years, Christianity has also made significant contributions to logic and science. Many of the advancements in the study of logic have come from the work of Christian-oriented scholars. In his logic textbook, *A Concise Introduction to Logic*,[14] contemporary logician Patrick Hurley lists ten "eminent logicians." Six of them have connections to historic Christianity: Peter Abelard (c. 1079–1142), William of Ockham (c. 1287–1347), Gottfried Leibniz (1646–1716), George Boole (1815–1864), John Venn (1834–1923), and Kurt Gödel (1906–1978). Christians have a high view of the capacity of human reasoning as grounded in the image of God and reflected through Christ as the "reason of God" (Greek: *logos*, meaning "word, discourse, or reason," see John 1:1). Historic Christianity prizes reason as a good gift from God.

Christianity played a decisive role in shaping the intellectual climate that gave rise to the Scientific Revolution in Europe.[15] Not only were most of the founding fathers of science devout Christians themselves,[16] but the biblical worldview provided a basis for modern science to emerge and flourish. Christian theism affirmed that an infinite, eternal, and personal God created the world *ex nihilo* ("from nothing"). The creation, reflecting the rational nature of the Creator, was therefore orderly and uniform. Furthermore, God created humans in his image (Genesis 1:26–28), making us uniquely capable of reasoning and of discovering the created order's intelligibility. In effect, the Christian worldview supported the underlying principles that made scientific inquiry possible *and* desirable.

Christianity and Reasonable Diversity
All branches of historic Christianity are generally united concerning the core truths of the faith (such as those expressed ecumenically in the Apostles' and Nicene Creeds). At the same time, the faith as a big tent allows for rigorous

rational debate and differing perspectives on secondary issues.[17] Naturally, disagreements among fallen humans can be taken too far. *All* groups struggle with infighting; Christianity is no exception. Nevertheless, it is often the case that Christian scholars provide a more thorough and vigorous treatment of controversial issues than is found in other religious traditions or within secularism. (For example, Christian scholars regularly debate controversial issues in philosophy, theology, and science from various sides, including the careful inspection of secular ideas and ideas from alternative religions.) The evangelistic and apologetic nature of the faith encourages a comparison of perspectives even within Christendom. Though this diversity of perspectives has sometimes led to divisiveness, it has also had a positive impact. In effect, historically speaking, Eastern Orthodoxy, Roman Catholicism, and Protestantism have competitively improved one another.

Faith Seeking Understanding in the Twenty-First Century
Many great modern-day Christian thinkers continue the broad "faith seeking understanding" tradition of historic Christianity, including (among many others) Richard Swinburne, Alvin Plantinga, Nicholas Wolterstorff, Marilyn McCord Adams, Peter van Inwagen, Eleonore Stump, Robert Audi, Joni Eareckson Tada, John Warwick Montgomery, William Lane Craig, Alexander Pruss, Joseph Ratzinger (aka Pope Benedict XVI), Peter Kreeft, Robert P. George, David Bentley Hart, N. T. Wright, Alister McGrath, Stephen T. Davis, Paul Copan, Gary Habermas, J. P. Moreland, Nancy Pearcey, Winfried Corduan, Krista Bontrager, Francis Collins, John Lennox, John Polkinghorne, Donald Page, James Tour, and Hugh Ross.

These various ways that Christianity relates faith to reason show the religion to be far from arbitrary and blind, but rather robustly grounded in knowledge and reason.

Christendom's A-Team

Christendom's high view of reason is also reflected among its greatest theologians and philosophers of the ancient and medieval past. From its beginnings, the Christian church has been graced with brilliant and influential thinkers. A list of "who's who" in Christian scholarship would be long and diverse in terms of scholarly disciplines. To whet your appetite in the areas of theology and philosophy, here's a snapshot of seven "straight A" thinkers. All of them were brilliant and accomplished scholars—no doubt, they would've received straight A report cards in today's grading system. These philosophers and theologians advanced Western civilization in general and Christianity in particular. It so happens that all their names begin with the letter *A*. So, they're Christendom's A-Team in a double sense![18]

Antony the Great of Egypt (c. 251–356) — Desert Monk

Antony was born in Egypt in a Christian family and had responded to his Christian calling by giving away all his possessions and retreating to the desert to pursue a monastic life of extreme asceticism. Antony organized his followers into a community of desert hermits. St. Athanasius's famous book *The Life of Antony* (AD 360) profiles Antony's life of spiritual warfare in which he battles the wiles of the Devil during his sojourn in the Libyan desert. Through various spiritual battles, Antony develops the spiritual disciplines of prayer, fasting, and solitude. Through Athanasius's account of Antony's extraordinary life of spiritual discipline, many people have considered a monastic vocation. When Athanasius's work was translated into Latin, it even influenced Augustine, who in his own work *Confessions* mentions Antony.

Amma Syncletica of Alexandria (c. 270–350) — Monastic Pioneer

Syncletica was the child of wealthy parents in Alexandria, Egypt. It's said that, due to her devotion to Christ, she gave her inheritance to the poor and left the city with her younger sister (who was blind) to live a life of chastity, poverty, and solitude. She eventually taught others who sought her out and provided guidelines for an early monastic order of women. These rules, recorded by her biographer (who some think may have been Athanasius), would later influence European monasticism. While Christendom has been traditionally patriarchal, there have been many scholarly women in its ranks through the centuries.

Athanasius (c. 296–373) — Defender of Orthodoxy

Athanasius is one of the most respected theologians in church history. All three branches of Christendom honor him. His articulation and defense of essential Christian doctrine earned him the title "father of orthodoxy." He was a rare combination of tenacious character and depth of theological insight. Athanasius championed the doctrine of the incarnation at a time when the faith was extremely vulnerable to heretical attack from Arianism (a fourth-century heresy that denied the deity of Christ). Athanasius met the intellectual and doctrinal critics of historic Christianity on their own grounds and successfully defended the faith. Two of Athanasius's most influential books are *On the Incarnation* and *The Life of Antony*.

Ambrose (c. 340–397) — The Great Orator
Ambrose is recognized as a doctor of the Catholic Church and is known as one of the great orators in Christian history. A classically educated scholar, he helped introduce Greek Christian thinkers to the Latin West. He was one of the leading Christian intellectuals in the Western church on par with some of the best Eastern church fathers. Theologically and apologetically, he successfully battled the Arian heresy and courageously stood up to the leaders of the Roman Empire during times of persecution against Christians. Ambrose was a skilled church theologian and served as the distinguished bishop of Milan. He was deeply instrumental in the conversion of St. Augustine and later baptized him into the faith. Ambrose was an influential writer, with significant contributions to the field of allegorical interpretation.

Augustine (354–430) — Greatest Author of Antiquity
Augustine of Hippo is arguably the most influential Christian thinker outside the New Testament authors—an intellectual giant. History knows him as a theologian, philosopher, church bishop, and a gifted and tenacious defender of Catholic Christianity. Augustine battled the Pelagian heresy (the view that one can be saved based on good works) and shaped such critical doctrines as original sin, creation *ex nihilo*, salvation by grace, and the Trinity. In apologetics, he developed a theodicy concerning evil and established the view that faith seeks understanding. His writings to some degree reflect a Christian-Platonic synthesis that explained and defended the Christian worldview through the prism of Plato's philosophical categories (more specifically through Neoplatonism). Having penned more than five million words, Augustine was the most prolific author of antiquity. He wrote several books that are considered both Christian and Western literary classics, including *Confessions* and *The City of God*. His other influential books include *On Christian Doctrine* and *On the Trinity*.

Anselm (1033–1109) — Integrator of Faith and Reason
Anselm of Canterbury is honored as a doctor of the Catholic Church and has been recognized as the greatest Christian thinker between Augustine and Thomas Aquinas. An astute theologian and philosopher, Anselm was one of the founders of medieval scholasticism that sought to show the rationality of Christian truth claims. He is well known for his understanding of the relationship between faith and reason and for formulating the ontological argument (as a maximally perfect being, God then must exist)—one of the most distinctive arguments ever for the existence of God. Anselm's book *Cur Deus Homo*

("Why Has God Become Man?") explains and defends the doctrine of the incarnation, and his work *Proslogion* includes his articulation of the ontological argument and other philosophical thoughts.

Aquinas (1225–1274) — The Catholic Philosopher Par Excellence
Thomas Aquinas may have possessed the brightest mind in the history of Christendom, if not in all Western civilization. A medieval scholastic philosopher and theologian, Aquinas developed a system of thought (called "Thomism") that was declared the official philosophy of the Roman Catholic Church. Aquinas also developed a Christian-Aristotelian synthesis that explained and defended the Christian worldview through the prism of Aristotle's philosophical categories. He argued that all language about God should be understood analogically. He is best known for his "Five Ways" of arguing for God's existence. In a life that spanned less than 50 years, he wrote voluminously and masterfully defended classical Christian theism. His two most important works are *Summa Theologica* and *Summa Contra Gentiles*.

As we saw earlier, some people today doubt whether the Christian faith is compatible with reason, but for much of Western civilization, the greatest intellectuals were also people of deep Christian faith. Today's Christians could greatly benefit from knowing about these scholars of their faith's ancient and medieval past.

Dealing with Doubts about the Faith
The rationality of Christianity doesn't eliminate entirely the various doubts people experience when it comes to belief in Christian truth claims. Several years ago, I received an email from a pastor who was going through some tough issues. He had battled a life-threatening illness and experienced difficult problems at his church. Worst of all, his daughter had begun having serious doubts about the truth of the Christian faith. This perfect storm of suffering had all struck at once. His letter really moved me. I have myself faced serious illness and I have a daughter the same age as the pastor's.

My response to the pastor focused on addressing the problem of doubt. Many Christians struggle with this issue at some point in their lives. In his excellent book *Dealing with Doubt*, apologist Gary Habermas identifies three types of doubt:[19]

1. **Factual doubt** – regarding the factual underpinnings of Christianity
2. **Emotional doubt** – stemming from subjective, psychological issues

(anxiety, depression, hurt, anger, grief, etc.)
3.   **Volitional doubt** – arising from a weak or immature faith

It seemed to me that the daughter's struggle with unbelief reflected all three types of doubt. Each type requires a different approach.

Factual doubt can be potentially relieved by reviewing solid apologetics material affirming the truth of the faith. It's best to choose resources that specifically address the facts causing doubt. For example, someone wondering about the reliability of the gospels might read my book *Without a Doubt*. Some of my other books focus on explaining the Christian worldview and comparing Christianity and Jesus himself to other world religions and their leaders (*A World of Difference* and *God among Sages*). In the case of the pastor's daughter, science had played a role in her doubt. Fortunately, my colleagues at Reasons to Believe have provided myriad science apologetics resources, like biochemist Fazale Rana's *The Cell's Design* and astrophysicist Hugh Ross's *Why the Universe Is the Way It Is*.

Emotional doubt requires a willingness to talk and listen—both from the person struggling with doubt and the person comforting and counseling them. The pastor's daughter was having a hard time coping with various trials the family had experienced in recent years. Spending time with trusted friends and bonding can help people open up to discuss the hurts, losses, and disappointments that they face. One of Christianity's greatest strengths as a world religion is the genuine hope and compassion it offers humanity. In his incarnation, Jesus Christ suffered both *for* us on the cross and *with* us in life's various disappointments. Jesus's family thought he was psychologically imbalanced (Mark 3:21). His disciples let him down when he needed their help the most, as he was preparing to confront crucifixion (Matthew 26:26–46). He experienced anger at the failures of the religious leaders of his day (Matthew 21:12–17). He grieved (John 11:17–44). He felt exhaustion (John 4:6) and stress (Luke 22:39–46). Hebrews 4:15 tells us, "For we do not have a high priest who is unable to empathize with our weaknesses, but we have one who has been tempted in every way, just as we are—yet he did not sin."

My correspondent's daughter had been hurt by people in the church. Many people—Christians, non-Christians, and former Christians—can empathize with her experience, including me. It is important for us to remember that Christians are forgiven sinners and that the brokenness and wounds of humans run deep—even in the church, unfortunately.

With volitional doubt, it is important to consider life stage, life events, and personality. The pastor's daughter was an adolescent, meaning she was at a critical stage in life where she was particularly vulnerable to doubt and confusion. When a person's faith is at a low ebb, the love shown by fellow Christians is itself a type of apologetic for Christianity. No other religion or worldview has agape love (doesn't expect anything in return).

As I know from personal experience, the trustworthiness of Scripture and the love of other Christians is a great help in dealing with pain and doubt. This is why it is important to invest in fields of study that establish and support Christianity's reliability. When I struggle with doubt, I remind myself of Jesus's promise in Matthew 11:28: "Come to me, all you who are weary and burdened, and I will give you rest." It is one of my very favorite passages of Scripture.

Christianity's Rational Explanation

Another powerful way to show the compatibility of Christianity and reason is to highlight the faith's genuine explanatory power and scope, which illustrate that the historic Christian faith and its accompanying worldview are both rational and true. In this regard, it's helpful to appeal to *best explanation* or *cumulative case* apologetics, which formally utilizes an abductive form of reasoning.

Most people recognize two basic forms of logical argumentation: deductive and inductive reasoning. But logicians sometimes speak of a lesser-known way of thinking called abductive reasoning. This third form of reasoning attempts to arrive at the best explanation for an event or series of facts. It seeks to provide the most plausible broad, explanatory hypothesis.[20]

A Cumulative Case for God

Just as a lawyer presents a brief in court or a criminalist lays out a forensic case or a physician arrives at a diagnosis by considering multiple symptoms and tests, a cumulative case consisting of multiple lines of converging evidence can be marshaled for the Christian worldview (see table 1.3). Consider Christian philosopher Richard Swinburne's assessment of this approach:

> Scientists, historians, and detectives observe data and proceed thence to some theory about what best explains the occurrence of these data. . . . We find that the view that there is a God *explains* everything we observe, not just some narrow range of data. . . . The very same criteria which scientists use to reach their own theories lead us to move beyond those theories to a creator God who sustains everything in existence.[21]

Table 1.3 Who Uses Abductive Reasoning?		
Diagnosticians: Moving from information to a coherent, plausible explanation		
Discipline	**Information**	**Explanation**
Detectives	Clues	Theory
Historians	Facts	Interpretation
Scientists	Data	Hypothesis
Physicians	Symptoms	Diagnosis
Mechanics	Problem	Solution

Christianity's ability to account for and justify the many diverse and undeniable realities of life and the world ranks as one of the strongest evidences that God exists and that the truth claims of the historic Christian worldview are both rational and correct. Christian theism (with its Creator and Redeemer God who is transcendent, infinitely intelligent, moral, personal, and loving) best accounts for the following realities far better than secular naturalism, which has no mindful or intelligent agent as a source.

The Cosmos: A transcendent Creator, not naturalism, accounts for a universe that has a singular beginning, shows design (order, regularity, and fine-tuning), and is susceptible to rational investigation.

Abstract Realities: An infinitely intelligent Creator, not naturalism, accounts for the existence and validity of the laws of logic, mathematical principles, and scientific models that correspond to the time-space universe as conceived in the mind of humans.

Moral Truths: A moral Creator, not naturalism, accounts for the existence of universal, objective, and prescriptive ethical values (goodness, justice), facts, and experiences.

Humans: A personal Creator, not naturalism, accounts for humankind's consciousness, rationality, free agency, relationality, enigmatic nature (greatness and wretchedness), moral and aesthetic impulse, and need for meaning and purpose in life.

Table 1.4 How Christian Theism Explains Our Reality

World
    1. Existence from a Creator
    2. Beginning from a causal power
    3. Design from an architect
    4. Investigation from a planner
    5. Abstractions from a thinker
 	
Humankind
    6. Consciousness from a greater consciousness
    7. Relationality from a person
    8. Morality from a moral being
    9. Purpose from a purposeful mind
  10. Religiosity from the divine
  11. Beauty from an artist
  12. Enigma from divine image and fall
  13. Eternity from the Eternal One
 	
Religious Revelation
  14. Incarnation from the Trinity
  15. Atonement by the God-man
  16. Resurrection from divine life
 	
Cumulative
  17. Successive additions

Religious Elements: A loving Creator, not naturalism, accounts for humankind's spiritual nature, religiosity, and religious experience, the miraculous events of Christianity, and the unique character, claims, and credentials of Jesus Christ.

These realities represent a cumulative case of compelling evidence for the God of the Bible. That is, while each of the individual points carries a certain logical or evidential force of its own, it's also true that the data taken collectively offers an even more formidable case in favor of the existence of the God of the Bible.

Other worldviews, both secular and religious, struggle to explain life's critical realities and sometimes even deny them. For example, some versions of naturalism deny consciousness, Eastern and postmodern views deny or diminish logic, some versions of Hinduism deny physical reality, and Buddhism denies the existence of the soul. By logical inference the God of Christian theism best accounts for and explains the meaningful array of realities encountered in the world and in life (see table 1.4).

The Ruling

While there are anti-intellectual Christians, historic Christianity as a whole clearly prizes reason. This respect for the rational sphere is seen in the way Scripture relates faith to reason and teaches Christians to be thoughtful people, in the great contributions that Christian scholars have made to rational fields of inquiry, in the number of brilliant scholars the faith has produced over the centuries, and in the profound rational explanatory power and scope that the Christian worldview wields.

Dawkins describes faith as "the great excuse to evade the need to think and evaluate evidence."[22] The evidence, however, demonstrates that this description of faith is inaccurate. Failure to acknowledge the depth and breadth of Christian rationality sets up a straw man that keeps people from engaging in profound discussion and debate.

Chapter 2

If God Created Everything, Then Who Created God?

by Kenneth Richard Samples

> If we are to understand the nature of reality, we have only two possible starting points: either the brute fact of the physical world or the brute fact of a divine will and purpose behind that physical world.
>
> —John Polkinghorne, "Serious Talk: Science and Religion in Dialogue"

Have you ever wondered where God came from? Or how God could be eternal? If you have asked these profound questions, you are not alone. Both Christians and non-Christians wrestle with God's eternal existence. When everything you know has a beginning, it is hard to come to grips with a being that has always existed. Yet the God of Christianity is qualitatively different in being from all other things. He is a necessary reality. In other words, God *must* exist. It's a deep concept to comprehend for humans who are temporal and finite by their very nature.

So, if God has created all things, does he also require a cause? This question is posed by small children, by college students, by leaders of atheist societies, and by secular philosophers and scientists. For example, in *A Brief History of Time*, Stephen Hawking asks, "Does it [the universe] need a creator, and, if so, does he have any other effect on the universe? And who created him?"[1] According to a 2015 *New York Times* article, "Who created God?" topped Google searches about God.[2]

"Who created God?" is essentially a question about the nature of causality. Answering this very common query effectively from a Christian theistic

---

Table 2.1 Two Distinct Cosmological Arguments

The Kalam cosmological argument and the contingency cosmological argument are two distinct arguments for God's existence. The first focuses on the need to explain the universe's specific beginning, whereas the second addresses how the universe, because it is a dependent reality, requires an independent reality.

Kalam Cosmological Argument
1. Whatever begins to exist has a cause for its coming into being.
2. The universe began to exist.
3. Therefore, the universe has a cause for its coming into being.

Contingency Cosmological Argument
1. All contingent realities depend on a noncontingent or necessary reality for their existence.
2. The universe is a contingent reality.
3. Therefore, the universe depends on a noncontingent or necessary reality for its existence.

---

viewpoint requires a combined philosophical, scientific, and theological response. To get a handle on this question, let's begin by examining causality as it relates to God from a philosophical perspective.

A Philosophical Reply

"If everything must have a cause, then God must have a cause,"[3] writes Bertrand Russell, eminent philosopher of the twentieth century. Many atheists share Russell's view. Richard Dawkins posits that "the designer hypothesis immediately raises the larger problem of who designed the designer."[4] Their argument is that if God *doesn't* need a cause, then the intelligent design claim that everything needs a cause is rendered false. And if God doesn't need a cause, then maybe the universe doesn't need one either. This reasoning sidesteps the basic cosmological arguments that have been around for a couple thousand years (see table 2.1).

The various cosmological arguments for the existence of God claim that all things in the world are contingent (meaning they depend on something else for their existence). Therefore, the very cosmos itself must depend on a being that exists independently or necessarily (a noncontingent entity).

The mistake that secularists like Russell, Dawkins, and Hawking make is believing that the various cosmological arguments insist that *everything* needs a cause. Yet this is not the case. Rather, sophisticated Christian theists argue that anything that *begins* must have a cause. An entity must be contingent to require a cause. Both cosmological arguments involve a critical philosophical distinction between a *contingent reality* and a *necessary reality*.

Contingent Being

A contingent being or reality is that which is caused (has a beginning), depends upon something else (is an effect), is finite in nature (reflects limitation or boundaries), and lacks an ultimate explanation in itself (can serve only as an intermediate explanation). A contingent reality could potentially either exist or not exist, but it could not bring itself into existence from sheer nothingness. Thus, a contingent reality requires something outside itself to cause its existence. So, in terms of explanation, a contingent entity requires a deeper causal accounting. The buck, so to speak, cannot stop with a mere contingent entity.

Necessary Being

On the other hand, a necessary being or reality is uncaused, independent, eternal, and ultimate. A necessary reality must exist—it cannot *not* exist. A necessary reality is self-explanatory. Moreover, a necessary being's own self-sufficient nature is the adequate reason for its own existence. So, in terms of explanation, the buck can indeed stop with a necessary entity. In other words, it would be logically superfluous and a category mistake to ask who created a necessary being (more on this later from a theological point of view). Table 2.2 contrasts contingent and necessary realities or beings.

So how does this philosophical distinction between contingent and necessary beings or realities help in answering the common question about God and causality?

Consider the universe itself. Big bang cosmology provides powerful evidence that the universe is contingent. According to the prevailing scientific view of cosmology that has held sway for about the last 50 years, the space-time-matter-energy universe had a distinct and singular beginning 13.8 billion years ago. The universe, therefore, appears to be an effect and is seemingly dependent upon something outside of and beyond itself (a transcendent causal

Table 2.2 Contrasting Contingent and Necessary Beings or Realities	
Contingent Being or Reality	**Necessary Being or Reality**
Caused	Uncaused
Begins	Beginningless
Dependent	Independent
Effect	First Cause
Finite	Infinite
Temporal	Eternal
Limited	Unlimited
Could be otherwise	Could not be otherwise
Intermediate explanation	Ultimate explanation
Buck can't stop here	Buck can stop here

agent). We'll note some of the specific scientific reasons for concluding that the universe began and look at a major alternative to the universe having a beginning a little later in this chapter. For now, it's important to remember that a contingent reality is sustained by something else and, thus, by definition cannot bring itself into existence. But since the universe came into existence (had a singular beginning), then some other reality must have caused or created it from nonexistence.

Furthermore, one contingent reality cannot fully explain another contingent reality. For example, atheist philosopher Daniel Dennett asks, "If God created and designed all these wonderful things, who created God? Supergod?

And who created Supergod? Or did God create himself?"[5] This line of thinking is not logically sufficient or coherent. When attempting to explain motion in the world, Aristotle cogently argued there must be a reality that causes but is itself uncaused (or, a being that moves but is itself unmoved). Why? Because if there is an infinite regression of causes, then everything is simply pushed back one step and then another step back and another. By definition, the whole process could never begin and, ultimately, nothing is truly explained (see table 2.3). Many Christian thinkers, like Thomas Aquinas (1225–1274), viewed Aristotle's reasoning on this point as logically probative. Here is Aquinas from the *Summa Theologica*:

> Therefore, whatever is in motion must be put in motion by another. If that by which it is put in motion be itself put in motion, then this also must needs be put in motion by another, and that by another again. But this cannot go on to infinity, because then there would be no first mover, and, consequently, no other mover; seeing that subsequent movers move only inasmuch as they are put in motion by the first mover; as the staff moves only because it is put in motion by the hand. Therefore it is necessary to arrive at a first mover, put in motion by no other; and this everyone understands to be God.[6]

In summary, then, the universe appears to be a contingent entity that can't stand on its own without a causal explanation. Christian scholars through the centuries have maintained that the contingent universe (a creation) requires a necessary reality (an eternally existent Creator) that by its own nature needs no causal explanation.

Causality

Another way of thinking about contingent and necessary realities is to consider two points about causality itself. The first is a fundamental principle of causality and the second is a logical inference from that principle.

The Latin phrase *ex nihilo nihil fit* ("from nothing, nothing comes") sums up the fundamental principle of causality. Everyday experience indicates that this principle is sound. Something cannot come from sheer or utter nothingness. For example, giraffes and Volkswagen Beetles don't just pop into existence out of thin air. No one has ever seen or experienced such an inconceivable thing. Even history's chief skeptic, philosopher David Hume, said, "I never asserted so

Table 2.3 Infinite Regress of Causes Fallacy
Universe → created by God → created by God^2 → created by God^3 → . . . But then how did this chain begin?

absurd a Proposition as that any thing might arise without a Cause."[7]

But what exactly is *nothing*? Well, it is literally *no thing*. Yet secular scientist Lawrence Krauss has proposed that a quantum fluctuation spawned the universe from nothing.[8] Krauss has been almost universally criticized for his use of the word *nothing*. For example, physicist Michael Strauss says Krauss's definition of *nothing* is clearly not nothing but rather something like "the underlying structure of our universe."[9] Reflecting upon sheer and utter nothingness will take your breath away.

So, what then is *nothing*? The specific definition of *nothing* is debated from both a philosophical and scientific perspective.[10] Perhaps the most provocative definition is attributed to Aristotle: "Nothing is what rocks dream about." Well, here's a pretty good definition for a start:

> Nothing = no matter, no energy, no space, no time, no quantum properties, no laws, no math, no logic, no mind, no consciousness, no reason, no potential, no actual, no cause, no effect . . .

A thing cannot come from nothing. A being cannot come from nonbeing. *Nothing* lacks the necessary potential to produce anything. As Christian philosopher Paul Copan rightly notes, "The chances of anything coming from absolute nothingness are zero."[11]

Undaunted, Hawking tried to sidestep this basic principle of causality by proposing the "exotic" idea that nothing really can produce something. He writes, "Because there is a law like gravity, the universe can and will create itself from nothing."[12] Hawking's brilliance in his academic field notwithstanding, this proposal can't be right. The law of gravity is clearly something—not nothing. Additionally, since the law of gravity is a principle of physics *within* the space-time universe, to say that gravity created the universe implies the incoherence that the universe existed in some form before it came into being. Even if Hawking's idea were correct, it would mean that the law of gravity is somehow *outside* of the space-time dimensions and is thus timeless, immaterial, and able to bring the universe into existence. In short, gravity begins to sound like the biblical God[13] that Hawking said doesn't exist. Again, the universe clearly cannot come from absolute nothingness.

"The fact of the matter is that the most reasonable belief is that we came from nothing, by nothing and for nothing," says atheist philosopher Quentin Smith. In the face of the reality that something cannot possibly come from utter nothingness, Smith believes we must "acknowledge our foundation in nothingness and feel awe at the marvelous fact that we have a chance to participate briefly in this incredible sunburst that interrupts without reason the reign of non-being."[14]

Take a moment and weigh the rationality of such a claim and belief. In *The Cambridge Companion to Atheism*, Smith gives a defense for what he calls a "Kalam Cosmological Argument for Atheism."[15] He accepts that the universe had a beginning, but he argues that the cause of that beginning shouldn't be identified as God. He contends that the universe is the cause of itself, but not in a self-contradictory way. It is important to note that some contemporary atheists recognize the challenge of something coming from nothing and think they have met the challenge with logical rigor. Yet, with all due respect, reason insists that something cannot come from sheer nothingness.

Philosopher Derek Parfit asserts that there is no question "more sublime" than "why there is a Universe."[16] This brings us to the logical inference from causality: if something exists and it cannot come from nothing, then there must be something that is eternal.

Similar profound reflections led Christian thinker J. Oliver Buswell to put it this way: "If anything does now exist, then something must be eternal, or else something comes from nothing without a cause."[17] In examining Buswell's claim, I think virtually everyone is confident that something does indeed exist (the world, ourselves, others). And since something cannot come from genuine nothingness, then it follows that something has always existed. Put another way, if sheer nothingness existed in the past, then nothing would exist right now (because something cannot come from utter nothingness)—but since something does indeed exist right now, it must be that something has always existed.

This eternal reality would not need a cause since only things that begin need a cause. So, the idea of an eternally existent, uncaused being is not logically problematic or even philosophically controversial. Some of the ancient philosophers, both Christian and non-Christian, believed in an eternal, uncaused, self-existent, and ultimate transcendental reality or being. Even some atheists who reject the existence of an eternal God nevertheless think the universe can be eternal and that it therefore doesn't require a cause. Theoretical physicist and atheist Sean Carroll does this very thing in stating, "Every attempt to answer the question 'Why is there something rather than nothing?' ultimately grounds in a brute fact, a feature of reality that has no further explanation."[18] For Carroll, the universe is an eternal brute reality without need of explanation. He states, "I think the question of 'Why is there something rather than nothing?' is interesting, but the answer probably is, 'That's just the way it is.'"[19]

But if the universe could be potentially eternal, then so could God. Christian philosopher William Lane Craig appropriately asks, "How can they [secularists] possibly maintain that the universe can be eternal and uncaused, yet God cannot be timeless and uncaused?"[20] The universe itself gives clear indication of being a contingent entity that by necessity requires a transcendent, uncaused cause.

A Scientific Reply

Philosopher Gottfried Leibniz (1646–1716) asked the ultimate metaphysical question: "Why is there something rather than nothing?"[21] Why indeed? And as we reasoned earlier, since something does exist and can't come from sheer nothingness, then doesn't that mean that there must have always been something in existence?

Some of the world's greatest thinkers have believed that the universe itself was eternal. Aristotle proposed this view, and the German Enlightenment philosopher Immanuel Kant (1724–1804) suggested the universe may be infinite in size and eternal in age. Prior to the middle of the twentieth century, even the modern scientific community entertained the idea of an eternal universe. Cosmological theories such as the steady state and oscillating models involved a universe that had always existed. These theories fell out of favor in the latter half of the twentieth century with the emergence of the big bang cosmological model that proposes a beginning to the cosmos.

The traditional big bang cosmological model, based on the observed expansion of the universe, provides strong evidence that the universe had a beginning,[22] marking it as a contingent entity. The big bang model asserts that the universe had a distinct and singular beginning 13.8 billion years ago. All matter, energy, time, and space exploded into existence (in a carefully controlled and fine-tuned manner) from nothing (no preexisting materials) and from an initial point of origin. This explosion generated astounding light and heat that gradually diffused and cooled to allow for the formation of stars and planets.

Here are brief summaries of some of the scientific reasons for concluding that the universe began:[23]

1. Albert Einstein's theorems of general relativity suggest an expanding universe. Expansion implies a beginning or point of origin. Einstein initially "buggered" his equations to get rid of the implied expansion, but later realized his blunder.

2. Astronomer Edwin Hubble subsequently observed that the universe is indeed expanding from a central point of origin with galaxies receding away from each other at incredible speeds.

3. Catholic priest and astronomer Georges Lemaître, using general relativity and the observed recession of galaxies, proposed a viable theory for the expansion of the universe, which implied a beginning.

4. The second law of thermodynamics involves a universe that is winding down (the amount of usable energy in the universe is continually decreasing over time), meaning the universe not only has a beginning but will have an end as well.

5. Given that the cosmos is winding down, if the universe were eternal and thus infinite in age, then the life cycles of stars would have already reached their completion and suffered heat death.

6. The discovery of the static noise or echo of the cosmic microwave background radiation that the big bang model had predicted would be left behind from the initial explosion serves to confirm the universe's beginning.

7. The oldest measured age of things observed in the universe clusters around 13.8 billion years, thus evidencing a temporal age for the cosmos.

This standard big bang cosmological model has held sway for the last 50 years. Most research scientists embrace it because it has withstood extensive scientific testing. This led Hawking and Roger Penrose to make the following statement about the singular beginning of the universe: "Almost everyone now believes that the universe, and time itself, had a beginning at the Big Bang."[24]

Following suit, leading astrophysicists John Barrow and Joseph Silk have stated, "Our new picture is more akin to the traditional metaphysical picture of creation out of nothing, for it predicts a definite beginning to events in time, indeed a definite beginning to time itself."[25]

Confirmation of the universe's beginning led agnostic philosopher

Anthony Kenny to remark, "A proponent of the big bang theory, at least if he is an atheist, must believe that matter came from nothing and by nothing."[26] But as we have reasoned, believing that something came from actual nothing and by nothing is not a rational viewpoint. The big bang carries with it heavy religious implications—namely that a beginning requires a Beginner.

The big bang universe, in total or its constituent parts, could have easily not existed. There is no reason to think the universe is anything other than a contingent reality. The universe, therefore, appears to be an effect and, thus, is seemingly dependent upon something outside of and beyond itself (a transcendent causal agent).

Rising Scientific Doubts about the Beginning of the Cosmos

Some scientists of late have begun to question whether the big bang represents the actual or specific *beginning* of all things. Questions about the universe's beginning are raised because cosmologists don't know what the universe was like at its earliest moments. Without a theory of quantum gravity, cosmologists don't know "what laws of physics governed the first 10^{-35} seconds or so of our universe or what laws, if any, existed beforehand."[27] Some cosmologists have conjectured that the earliest moments of the universe may not require a beginning but may allow for some kind of infinite past.

So, did the universe have a beginning or not? Michael Strauss provides the following insight:

> All the observational data we have extrapolates to less than a trillionth of a second after our universe started to expand and indicates the universe seemed to have an actual beginning. All theoretical calculations indicate the universe had a real beginning . . . Everything measured, observed, and calculated that has been confirmed points to a definite beginning. You can call it the big bang or whatever you want, but the data leads to a beginning. Of course, we don't know beyond any doubt what happened at the beginning or what caused the beginning. We do know beyond any doubt that everything is consistent with the biblical record that proclaims this universe had a beginning and that all the scientific evidence is in accord with the predictions based on the Bible.[28]

But if the data supports the universe having had a beginning, why are so many cosmologists seemingly doubtful about the cosmos having begun? Again, Strauss offers his assessment:

> Everything that we do know about the origin of our universe seems to indicate it had an actual transcendent cause consistent with the Christian God. Those who want to remove God from the equation must appeal to what is not known, rather than to what is known. I call this an "atheism of the gaps" . . . It is ironic that atheists have for years claimed that Christians appeal to a god of the gaps to explain things that are not known, but many of the current arguments from atheists against God can only appeal to gaps in our understanding.[29]

To be fair, Christian scholars who are critical of biological evolution often appeal to gaps in scientific understanding concerning the biological origin of humans. It seems gaps will always give rise to questions and to possible alternative interpretations of the scientific consensus on issues. So, these novel interpretations will have to be scrutinized and accepted only if they provide genuine explanatory power and scope.

Multiverse Musings

Many within the scientific community now consider a multiverse[30] (also referred to as the many-worlds hypothesis[31]) as a potential explanation for the existence of our universe. According to multiverse theories, our universe is just one of many, a tiny region of an unfathomably vast multiverse produced by some cause or mechanism outside the known cosmos. Some multiverse proponents speculate that our universe may be the only one with the exacting parameters required for the emergence of complex life-forms.

It is possible that the multiverse idea has some basis in highly speculative (yet to be verified) mathematics. Nevertheless, it must overcome some serious challenges. Let's consider seven specific obstacles.[32]

First, multiverse theories are currently unverifiable and unfalsifiable, causing many scientists to deem them speculation—not science. As of now, no empirical data directly supports the existence of other universes. And the fact that hypothetical other universes would exist outside this one may mean that detecting them would never be possible. Physicist George Ellis made this stinging comment:

> Parallel universes may or may not exist; the case is unproved. We are going to have to live with that uncertainty. Nothing is wrong with scientifically based philosophical speculation, which is what multiverse proposals are. But we should name it for what it is.[33]

Second, secular-oriented scientists need to consider whether this theory violates the Ockham's razor principle with overly complex or conveniently makeshift (*ad hoc*) explanations of the data.

Third, assuming many universes exist may result in a type of infinite regression fallacy. So how did the multiverse begin and what is its ultimate origin? If there are "metalaws" outside the multiverse that brought it into existence, where did they come from? It is not logically acceptable to merely push the explanation back one more causal step.

Fourth, on an existential level, are secular scientists willing to go all in on an unobservable and unverified (probably even unverifiable) theory when they criticize Christians for considering Pascal's wager (the cost-benefit analysis of accepting or rejecting potential states of affairs after death)?[34] Cosmological theories have philosophical and possibly theological implications for the human condition and for what potentially awaits people after death.

Fifth, the multiverse seems to imply the existence of something supernatural or above the natural realm of reality. Can a theory dependent on forces outside space-time be considered naturalistic? Appealing to mechanisms rather than empirical data clashes with old-school atheistic naturalism (physicalism or materialism).

Sixth, some of the leading proponents of multiverse theories have accepted the idea that an expanding universe must have had a beginning.[35] For example, after examining theories that claim to avoid a need for a beginning to the universe, cosmologist Alex Vilenkin has said:

> "None of these scenarios can actually be past-eternal," and "All the evidence we have says that the universe had a beginning."[36]

Seventh, even if a multiverse model emerges as viably correct, the view in general could still be compatible with Christian theism. Scripture teaches that God created at least one other realm of existence besides the space-time universe (the angelic realm; see Colossians 1:16).

Astrophysicist Jeff Zweerink comments:

Table 2.4 Two More Cosmological Arguments

Dependence Cosmological Argument
1. Every dependent thing needs something to depend upon.
2. The universe (the sum of material reality) is a dependent thing.
3. Therefore, the universe needs something to depend upon (God).

Leibnizian Cosmological Argument
1. Everything that exists has an explanation of its existence, either in the necessity of its own nature or in an external cause.
2. The universe exists.
3. Therefore, the universe has an explanation of its existence.
4. If the universe has an explanation of its existence, that explanation is God.
5. Therefore, the explanation of the universe's existence is God.

Rather than get stuck on whether we live in a multiverse, I prefer to focus on the more important question: Does a multiverse fit more comfortably in the naturalist's worldview or the Christian's? Upon my first exposure to the multiverse, I thought it was a threat to Christianity. However, further investigation revealed that multiverse research strengthens the case for a beginning and for design—and thus a Beginner and Designer.[37]

Lest one think that the multiverse has a scientific leg up on a theological interpretation, consider the words of agnostic physicist Paul Davies:

Extreme multiverse explanations are therefore reminiscent of theological discussions. Indeed, invoking an infinity of unseen universes to explain the unusual features of the one we do see is just as ad hoc as invoking an unseen Creator. The multiverse theory may be dressed up in scientific language, but in essence it requires the same leap of faith.[38]

Christian philosopher Richard Swinburne thinks invoking a near-infinite number of universes to explain what a single God can do in one universe isn't reasonable.[39] He says, "It is crazy to postulate a trillion (causally unconnected) universes to explain the features of one universe, when postulating one entity (God) will do the job."[40]

It seems that the multiverse view is unable to sidestep philosophical and scientific questions about the need for a beginning and thus a cause. As a highly speculative theory that can't be tested empirically, the multiverse view does not stand as a serious obstacle to the viewpoint that the universe had a beginning and thus is dependent upon a Beginner (see table 2.4).

So, the traditional big bang cosmological view continues to point toward a contingent universe that is dependent upon something else for its existence. The universe appears to be winding down and gives no indication of being self-sufficient or necessary.

In modern cosmology, the creation implication remains powerful. If scientists follow the observational data and resist speculating upon the unknown aspects of the earliest moments of the cosmos, then the universe appears to have had a singular beginning a finite time ago. This standard big bang cosmological model uniquely corresponds to the biblical teaching of creation *ex nihilo*. Arno Penzias, the Nobel Prize–winning physicist who discovered the cosmic background radiation, declared, "The best data we have are exactly what I would have predicted, had I had nothing to go on but the five books of Moses, the Psalms, the Bible as a whole."[41] The Scriptures, written a couple thousand years ago, contain a view of cosmology that corresponds to the most plausible scientific findings. The Bible opens with the declaration that "in the beginning God created the heavens and the earth" (Genesis 1:1).[42] And the psalmist contrasts the temporal cosmos with the eternal Creator: "Before the mountains were born or you brought forth the whole world, from everlasting to everlasting you are God" (Psalm 90:2).

A Theological Reply

A theological understanding of God's nature can help address the question as well. According to the Bible, God is self-existent. Theologians refer to this trait or property as God's attribute of aseity. God does not require, nor does he rely upon, anything outside of himself (such as the created order) for his perpetual existence. When the apostle Paul spoke to Greek philosophers in Athens, he contrasted the biblical God with the traditional polytheistic view of deities:

The God who made the world and everything in it is the Lord of heaven and earth and does not live in temples built by human hands. And he is not served by human hands, as if he needed anything. Rather, he himself gives everyone life and breath and everything else. . . . "For in him we live and move and have our being" (Acts 17:24–25, 28a).

Aseity makes the biblical God qualitatively different from all creation. Theologian J. I. Packer puts it this way: "He [God] exists in a different way from us: we, his creatures, exist in a dependent, derived, finite, fragile way, but our Maker exists in an eternal, self-sustaining, necessary way."[43]

God as an uncaused, self-sustaining, eternal, ultimate, and necessary being is the final explanation for all contingent realities. The prophet Isaiah illustrates God's aseity with a series of provocative questions (Isaiah 40:13–14):

Who can fathom the Spirit of the Lord, or instruct the Lord as his counselor? Whom did the Lord consult to enlighten him, and who taught him the right way? Who was it that taught him knowledge, or showed him the path of understanding?

God being triune also means that he is complete in himself—he does not require our love or companionship. And God as a personal being is also unlike mere abstract entities that have no power to produce something causally. God exercises his will to create all things.

John's Gospel identifies Jesus Christ as being a divine coagent in creation:

In the beginning was the Word, and the Word was with God, and the Word was God. He was with God in the beginning. Through him all things were made; without him nothing was made that has been made (John 1:1–3).

Jesus, the second person of the Trinity, declares, "I am the Alpha and the Omega, the First and the Last, the Beginning and the End" (Revelation 22:13). Philosophy, logic, science, and theology bear out the truth of this claim. The triune God's aseity not only makes him the best explanation for why there is something rather than nothing, it also makes him the only being worthy "to receive glory and honor and power" (Revelation 4:11, NIV 1984).

Therefore, the God of the Bible reveals himself to be an eternal and

Table 2.5 God's Nature

The triune God of historic Christianity is:

• Uncaused	• Atemporal	• Ultimate
• Self-sustaining	• Nonspatial	• Necessary
• Eternal	• Immaterial	• Personal

self-sufficient being without beginning or end (see table 2.5). God is the logically necessary being that explains why all the contingent realities of the universe have actual existence.

The Ruling

From a Christian perspective, to ask who created God is to engage in a category error. It is like asking what the musical note D smells like. God is in a distinct category from the rest of being and reality. An eternal, uncaused, self-existent, and ultimate transcendental being either requires no causal explanation or the explanation lies within this being's very nature. In other words, no one created God because God isn't the sort of being that requires or needs to be created. God stands as "unmakeable." He has always existed and always will. As Christian philosopher Paul Copan notes, "A caused God is no God at all, since He would be dependent on something else for His existence."[44]

Chapter 3

Creation of the Cosmos
by Hugh Ross

"In the beginning God created the heavens and the earth" (Genesis 1:1). With this simple yet profound declaration, the biblical account of God's plan for humanity begins. Thousands of pages of commentary have been devoted to this one statement alone.[1] Its explosive impact bursts upon the reader like the creative blast physicists have come to call the "big bang."

The assertion of a *beginning* immediately catches our attention. For centuries the philosophical pendulum has swung back and forth on the questions of the eternality of matter, energy, space, and time. Immanuel Kant was neither the first nor the last but perhaps the most convincing to propose an infinitely-old universe model.[2] His concept donned scientific garb as the "steady state" model.[3] Later still, scientists revived the Hindu doctrine of a universe that oscillates forever through cycles of birth, death, and rebirth.[4] However, the Bible says in unequivocal terms that the "heavens and the earth" *began,* that they exist for a finite time only,[5] and that God exists and acts inside, outside, and beyond the universe's space-and-time boundaries.[6] He alone is everywhere present and always existing.

The First Act

The Hebrew verb for "created" (*bārā'*) in its basic form (used here) appears in the Bible with only one subject: God. Its usage suggests the kind of creating that only God, and no one else, can do. Old Testament and Semitic languages scholar Thomas E. McComiskey comments: "This distinctive use of the word is especially appropriate to the concept of creation by divine fiat."[7] He adds that this verb choice "denotes the concept of 'initiating something new'"[8] and that "since the primary emphasis of the word is on the newness of the created object, the word lends itself well to the concept of creation *ex nihilo* [out of nothing]."[9]

Creation out of nothing can mean different things in different contexts (see sidebar, "Nine Kinds of Nothing"). So we turn to the creation statement in Hebrews 11:3 for clarification of the intended meaning: "The universe was formed at God's command, so that what is seen was not made out of what was visible." Astronomers can see or detect matter, energy, space, and time (or their effects) throughout the cosmos. Thus, according to Hebrews 11:3 (among other Bible passages[10]) God operated outside, beyond, or transcendent to matter, energy, space, and time when he created our universe of matter, energy, space, and time.

The verb *bārā'* appears just twice more in Genesis 1, in contrast to the frequently used verbs *'āśâ* (make, fashion, execute, manufacture),[11] *hāyâ* (be, happen, come about),[12] *dāshā'* (sprout, bring forth, or flourish),[13] *nātan* (set, put, place, give, or appoint),[14] *rā'â* (be seen, reveal, cause to be seen, or be shown),[15] and *yāṣā'* (go out from, come out from, bring forth, produce, or spring forth).[16] These two other uses of *bārā'* (verses 21 and 27) would also seem to imply that God brought into existence something new, something that did not previously exist.

The Heavens and the Earth

Hebrew differs significantly from English in many respects, including its vocabulary size. While English words number in the hundreds of thousands (depending on how one counts them), biblical Hebrew is comprised of a few thousand words.[17] To understand the meaning of *shāmayim* and *'ereṣ* ("heavens" and "earth") requires more than knowing the definition of each individual term.

'Ereṣ has six different meanings: the soil; the territory or land possessed by an individual, family, tribe, or nation; a city state; the territories of all peoples and nations; the underworld; or all the land and water, as well as the foundations that support them (what we now know as the planet Earth).[18] *Shāmayim*, a plural form (hence, "heavens"), has three meanings: the part of Earth's atmosphere where rain clouds form, that is, the troposphere; the abode of the stars and galaxies; and the spirit realm from which God rules.[19] New Testament writers and ancient and modern rabbis sometimes used the ordinals "first," "second," and "third" to identify which of these "heavens" they meant.[20]

According to Old Testament scholar Bruce Waltke, the phrase *hashamayim we ha'ereṣ* ("heavens" plural combined with "earth" singular)

Nine Kinds of Nothing

Scientists, philosophers, and theologians use different definitions of *nothing* depending on the context. Nothing can mean the complete absence of:

1. matter;
2. matter and energy;
3. matter, energy, and the three large cosmic space dimensions (length, width, and height);
4. matter, energy, and all the cosmic space dimensions (including the six extremely tiny space dimensions implied by particle creation models);
5. matter, energy, and all the cosmic space and time dimensions;
6. matter, energy, cosmic space and time dimensions, and created nonphysical entities;
7. matter, energy, cosmic space and time dimensions, created nonphysical entities, and other dimensions of space and time;
8. matter, energy, cosmic space and time dimensions, created nonphysical entities, and other dimensions or realms—spatial, temporal, or otherwise; or
9. anything and everything real, created and uncreated.

Because God is an eternal Being, without beginning or ending, only the first eight kinds of nothing in the list above are possible. As for the universe, it came from nothing, as in definitions 5, 6, or 7 above.

carries a distinct meaning, as would a compound noun in English. For example, when we put together "under" and "statement" or "dragon" and "fly" to form a compound noun, these words take on a specific and distinct definition. Similarly, the Hebrew expression *hashamayim we ha'eres* refers uniquely to the totality of the physical universe, all its matter, energy, space, and time.[21] Biblical Hebrew includes no single word for the now familiar concept of a universe. Michael Shermer is mistaken in presuming that Genesis 1:1 refers to the universe and to planet Earth separately.

A Unique Doctrine Validated

According to Genesis 1:1, the entire universe came into existence, entirely new, a finite time ago, by the creative act of God. This statement reverberates throughout the pages of Scripture.[22] No other "holy book" makes such a claim on its own. The concept appears elsewhere only in those books that borrow from the Bible, such as the Qur'an and the Mormon writings.

By contrast, those sacred books with no clear connection to the Bible claim that a god, gods, or forces created the cosmos *within* space and time, which, they say, have always existed. The Bible stands apart in declaring that space and time are not eternal but, rather, suddenly came into being by an act of God, a Being completely independent from—that is, transcendent to or outside of—space, time, matter, and energy.

The importance of this unique doctrine cannot be overstated. It sets biblical revelation apart from all other so-called revelatory writings. In light of what science has discovered, it provides potent evidence for the supernatural accuracy of Genesis.

Modern scientific support for the Genesis 1:1 creation event first arose from observations of the recession velocities of galaxies, which indicated a cosmic beginning. Rigorous verification first came via a theorem published by Stephen Hawking and Roger Penrose in 1970.[23] Their work on the space-time theorem established that a universe containing mass and in which general relativity reliably describes the motions of astronomical bodies *must* be traceable back to a beginning of space and time, implying it was brought into existence by a causal Agent who transcends space and time.

Over a ten-year span following the development of this theorem, researchers Arvind Borde, Alan Guth, and Alexander Vilenkin published five extensions arising from it. These extensions culminated in a theoretical "proof" showing that any *reasonable* cosmological model, that is, any model in which the universe expands (on average) throughout its history (the only conceivable cosmological models that will permit the existence of physical life), requires an actual beginning of space and time, implying the cosmos was initiated by a causal Agent operating beyond space and time.[24]

Turning Point

Something happens between verses 1 and 2 that powerfully impacts the reader's comprehension of the story to follow. Here, the frame of reference for the creation account shifts from the entire cosmos (the heavenly objects that make up the universe) explicitly to the surface of Earth. Perhaps

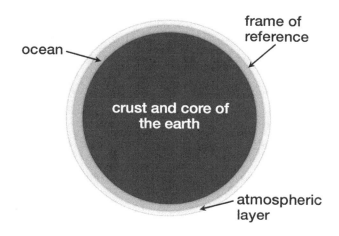

Figure 3.1: Frame of Reference for Genesis 1:3–31
The events of the six creation days are described from the vantage point of Earth's primordial (water-covered) surface—underneath the cloud layer, as Genesis 1:2 indicates—and its early inhabitability for life. *Diagram Credit: Reasons to Believe*

because it comes so abruptly, this transition is easily missed, even by distinguished Bible scholars. I am persuaded that my immersion in science prepared me to see it. In fact, I was struck with amazement that this ancient document would be structured much like a modern research report.

The same steps scientists use to analyze and interpret natural phenomena appear on the Bible's first page. At the time I was unaware, as many people still are, that the step-by-step process we now know as the scientific method owes its formation to individuals familiar with the Bible. They recognized a pattern in biblical texts that describe a sequence of events (see sidebar, "Biblical Testing Method/Scientific Method"). The Genesis account, for example, clearly identifies the frame of reference (or viewpoint) from which the sequence of events is described, including a statement of the initial conditions, the chronology, the final conditions, and some conclusions about what transpired. Within the Bible itself we see instructions to consider contextual elements essential for developing correct interpretations.[25] We also see warnings against the dangers of overlooking them.[26]

Biblical Testing Method/Scientific Method

Although the wording and number of steps delineated may vary slightly from one introductory science text to another, the basic components of the scientific method include these seven:

1. Identify the frame(s) of reference or point(s) of view.
2. Determine the initial conditions.
3. Perform an experiment or observe the phenomenon, noting what takes place when, where, and in what order.
4. Note the final conditions.
5. Form a hypothesis about the how and the why of the phenomenon.
6. Test the hypothesis with further experiments or observations.
7. Revise the hypothesis accordingly.

These steps apply just as strategically to biblical interpretation as to nature studies. In the biblical context, step 3 calls for noting all the explanatory and descriptive details, and step 6 calls for testing the initial understanding (hypothesis) with parallel and/or relevant biblical passages. This approach may not guarantee objectivity and accuracy, but it certainly helps minimize the effects of oversight, personal bias, and presuppositions. Nevertheless, all students of science and Scripture must recognize that our knowledge and our comprehension of that knowledge remain limited. So, our interpretations always fall short of perfection, and we must be willing to adjust and fine-tune them as research continues. Conclusions must remain open to ongoing testing and refinement as we repeatedly apply the biblical testing method.

The advance of science over the past few centuries may be attributed, in large measure, to rigorous and repeated application of this step-by-step method. It keeps us moving closer and closer to a correct understanding of the natural realm. If and when students of Scripture apply this systematic approach just as rigorously to biblical interpretation, it moves us closer to "rightly dividing the word of truth." It promotes unity rather than conflict in establishing sound doctrine. In the case of the Genesis creation and flood accounts, applying the scientific method—derived primarily from these very portions of Scripture—offers our best hope for developing a consistent interpretation, one free of both internal and external contradiction.

From the Heavens to Earth's Surface

These words depict the shift of perspective in Genesis 1:2:

> Now the earth was formless and empty, darkness was over the surface of the deep, and the Spirit of God was hovering over the waters.

The observer's vantage point is clearly identified as "the surface of the deep ... over the waters." Yet the vast majority of Genesis commentaries mistakenly proceed as if it were still high in the heavens, somewhere in the starry realm above Earth. This one oversight seems to account for more misunderstanding, more attacks on the credibility of Genesis, than all other interpretive errors combined. The problem glares from the page at anyone slightly aware of how nature works. If the storyteller's viewpoint lies in the heavens above, the unfolding sequence of creative events contradicts the scientific record. It violates much of well-established Earth (and life) history. For example, it would place the production of plants *before* the formation of the Sun, Moon, and stars.

Initial Conditions

In addition to clarifying the point of view, the Genesis 1 creation account also identifies four of Earth's major features at the outset of the narrative. One of these conditions is darkness. As the creation days begin, the darkness is pervasive. A look ahead to the third creation day reveals that water initially covered Earth's entire surface (Psalm 104:6):

> You [God] covered it [the earth] with the deep as with a garment;
> the waters stood above the mountains.

The book of Job makes reference to both the darkness and the water. Job 38:9 says that Earth's surface was dark because of opaque enshrouding clouds. God says in reference to "the sea" that covered Earth's surface, "I made the clouds its garment and wrapped it in thick darkness."

According to the Genesis account, no continents initially rose above the water, and the whole of Earth's watery surface remained in darkness. No visible light reached through Earth's primordial atmosphere.

Next, the passage mentions two more conditions, *tōhû wābōhû*, translated variously as "formless and empty;" "without form and void," and "unformed and unfilled." Given that the creation account focuses on God preparing Earth for life and filling it with life, the reference to "unformed" for life and "empty" of life makes sense. As Old Testament scholar David Tsumura explains, *tōhû wābōhû* is not a description of chaos, but instead refers to Earth as "an unproductive and uninhabited place."[27]

Context holds the key to interpreting *tōhû wābōhû* (see sidebar, "What about the Gap Theory?"). The story builds step by step toward the climactic moment when God created humanity. Thus, each creative act highlighted bears significance in relation to God's preparations and provisions for humans' arrival and sustenance.

Completed Acts

In analyzing the structure and grammar of Genesis 1:1 and 1:2, Hebrew linguists have determined that those texts proclaim that the creation of the universe and the formation of Earth must predate the events described in the six creation days by an unspecified but finite duration of time. Consequently, whichever one of the four usages (see sidebar, "How Long Are the Creation Days?") one might choose for the Hebrew noun *yôm* (translated "day") to delineate the duration of the six creation days, Genesis 1 allows for both the universe and Earth to be as old as astronomers' and physicists' measurements have determined.

In his book, *Genesis 1–4: A Linguistic, Literary, and Theological Commentary*, C. John Collins writes, "The verb *created* in Genesis 1:1 is in the perfect, and the normal use of the perfect at the very beginning of a pericope [an excerpt from a text] is to denote an event that took place before the storyline gets under way."[28] He adds, "A number [of narrative pericopes] do begin with a verb in the perfect, and they do so in order to describe an event that precedes the main storyline."[29]

Rodney Whitefield in his book, *Reading Genesis One: Comparing Biblical*

Hebrew with English Translation, and his booklet summary, *Genesis One and the Age of the Earth,* explains that biblical Hebrew verbs by themselves do not specify the duration of actions. Nor do they determine the time ordering of actions. Instead, the ordering of past actions is established most straightforwardly by word order.[30]

Whitefield points out that most frequently in biblical Hebrew narrative the verb appears first followed by the subject of the verb. This is the case for the verbs that appear in all but three of the thirty-one verses that comprise Genesis 1. Exceptions, however, are found in Genesis 1:1 and 1:2. In describing the creation of the universe and the status of Earth, Genesis 1:1 and the first clause in Genesis 1:2, respectively, place the verb in the second rather than the first position. Placing the verb as the second word of a sentence or clause is one way, Whitefield explains, that biblical Hebrew establishes that a particular action has already been completed.[31] Thus, Genesis 1:1 declares that the universe has a beginning and that its creation is a completed event. Genesis 1:2 proclaims that Earth had existed in a formless and empty state. That is, the universe and Earth already are in place before the events of the six creation days.

The phrase "and God said" that (starting in Genesis 1:3) leads off the eight creative commands in the account of the six creation days confirms the conclusion of Collins and Whitefield. In each case, the eight such occurrences of "and God said . . ." starts a sequential narrative that follows up on what transpired before. Therefore, just as the text implies that the events of creation day three, for example, occurred after the events of creation day two, so also the events of the first creation day must have taken place after the events described in Genesis 1:1 and 1:2.

The completed nature of the creation of the universe and the formation of the primordial Earth implies that an unspecified duration of time transpired between the creation of the universe and the formation of Earth. Likewise, an unspecified time period transpired between Earth's initial formation and the events of creation day one.

Initial Conditions Confirmed

It's worth noting that water is one of the most abundant molecules in the universe. So, the watery covering of primordial Earth is no surprise. What ongoing research has revealed, however, is that a remarkable, exactingly orchestrated event adjusted the quantity, states, and distribution of Earth's water and atmosphere in a manner uniquely suitable for land-dwelling

What about the Gap Theory?

A century ago, when geologists were first discovering the long geological time spans, some biblical scholars put forth the gap theory, which interprets the *tōhû wābōhû* of verse 2 as implying a significant time gap in the account of creation.[32] They take the Hebrew verb *hāyâ* (translated "was") to mean "became." In other words, they suggest that Earth did not *begin* formless and empty but rather *became* formless and empty.

One rationale for this translation arises from the observation that *tōhû wābōhû* tends to carry a negative or pejorative connotation elsewhere in the Bible. A second rationale comes from the desire to reconcile the voluminous scientific evidence for Earth's antiquity with the popular twentieth century teaching that the Genesis days represent six consecutive 24-hour periods.

What's called the "gap theory" proposes that the beautiful world God created "in the beginning" suffered ruination (most often attributed to the actions of Satan and his fellow rebellious angels) and that the six-day creation account in Genesis 1 describes its restoration. According to this theory, astronomers, geophysicists, and paleontologists are measuring the ancient, ruined creation, whereas the Bible addresses God's recent repair of creation.

This theory, popularized by a C. I. Scofield study Bible published at the beginning of the twentieth century,[33] still holds sway among some Bible interpreters.[34] However, it falters on several significant points, both biblical and scientific. Perhaps most significantly, the Hebrew conjunction *waw* in Genesis 1:2 is not connected with the verb *hāyâ* in a manner that would mean "became." Genesis 2:7 (man became a living being) has the *waw*-consecutive with *hāyâ* such that the definition "became" is implied. However, in Genesis 1:2, *hāyâ* is in the perfect form without an immediately preceding *waw*. This construct eliminates the possibility of translating *hāyâ* as "became" rather than "was."

For a more thorough, in-depth discussion of the gap theory and its flaws, see *The Christian View of Science and Scripture* by theologian and philosopher of science Bernard Ramm.[35]

creatures, including humans. More of that story appears in the account of the Moon's formation.[40]

Through ongoing research into how planets form, scientists have been able to explain and confirm not only the dark and watery but also the "formless and empty" features of early Earth. Technological advances in the 1970s allowed astronomers to observe "disks" around young stellar objects. Thousands of these objects have been studied thus far, and each one is surrounded by an extensive disk of gas, dust, and debris.[36] Theoretical studies as well as observational evidence show how these disks eventually condense into planetary systems.[37] Indeed, such planetary systems are being discovered at a rapid rate. To date, astronomers have discovered more than 5,000 planets (or likely planets) and have measured the physical and orbital properties of more than 3,000.[38]

They've learned that planets as massive as Earth and as distant from their host star (their "sun") typically start with a thick, opaque (light-blocking) atmosphere. The smallest of the extrasolar (outside our solar system) planets for which astronomers have a measurement of the planet's atmospheric mass is 6.5 times more massive than Earth and has an atmosphere at least 4,000 times "heavier" than Earth's atmosphere today.[39] Venus, which measures 19 percent less massive than Earth (implying that its weaker gravity will be less able to accrete an atmosphere) and 28 percent closer to the Sun (implying that its greater planetary surface temperature will cause more of its atmosphere to dissipate to outer space), nevertheless possesses an atmosphere 91 times more massive than Earth's. Thus, astronomers estimate that Earth's primordial atmosphere was at least 200 times more massive than our current atmosphere. So visible light from the Sun (or stars and other heavenly objects) would have been unable to penetrate to the early Earth's surface.

No life-supporting landmasses would have existed on early Earth either. Initially all the land in Earth's rocky crust lay below the surface of the deep. Islands and continents arose gradually—think hundreds of millions of years—as a result of volcanism (volcanic activity) and plate tectonics (movement and collisions of large crustal sections). Volcanism and plate tectonics, driven primarily by heat from radioisotope decay in Earth's mantle, generated the wrinkling of Earth's surface. This wrinkling, which eventually pushed land upward above the ocean's surface, continues to this day, but at a much lower rate.[40]

Tectonic and volcanic activity superseded erosion (the wearing down

---

How Long Are the Creation Days?

In contrast to English the vocabulary size in biblical Hebrew is tiny. If one discounts the names of people and places, biblical Hebrew contains only about three thousand words.[41] Consequently, most nouns in biblical Hebrew possess multiple "literal definitions" or common usages.

The Hebrew noun, *yôm*, translated "day" in Genesis 1 is no exception. It has four distinct literal definitions:[42]

1. part of the daylight hours; for example, from noon to 3 PM
2. all of the daylight hours
3. a 24-hour period
4. a long but finite time period

While modern-day Hebrew has two words for an extended, finite-duration time period, in biblical Hebrew no other word besides *yôm* possesses the meaning of a long but finite period of time.[43] Therefore, if Moses wanted to communicate a creation history consisting of six eons, he would have no other option but to use the word *yôm* to describe those eras.

With many distinct literal definitions for so many of the Hebrew nouns, how does the reader determine which ones apply in a specific biblical text? The answer lies in the grammar, sentence structure, and context.

---

processes) until landmasses rose up above the oceans to cover about 29 percent of Earth's surface. They still do exceed the erosion rate, but to a much lesser degree. (Note: Earth's rotation rate has decreased by a factor of three or more over the past four billion years as a result of tidal interactions among Earth, the Sun, and the Moon.)[44]

Observational data, including the study of Earth's oldest rocks, and theoretical modeling of planetary formation together verify the historical accuracy of early Earth's conditions described in Genesis 1:2. Water did, indeed, initially cover all of Earth's crust.[45] The "formless" and "empty" conditions of Earth—relative to life—would appropriately depict a roiling surface shrouded in darkness by an opaque atmosphere and an accompanying cloud of interplanetary debris. Without light on Earth's surface, photosynthesis could not occur. With large pieces of interplanetary debris crashing onto Earth's surface at that time, no surface life could have survived. We can easily understand why we find no evidence of life on Earth prior to 3.8 billion years ago.[46]

The description of these conditions dramatically—and compellingly—sets the stage for the chronology of divinely engineered transformation and creation events presented in Genesis 1:3–27, the "days" of creation.

Chapter 4

Why Such a Vast Universe?
by Hugh Ross

The sheer enormity of the universe is enough to make anyone feel inconsequential. This feeling raises questions: Does life really have any ultimate value, meaning, or purpose? If God is responsible for our existence, why would the universe be so large?

Although skeptics once argued that the universe was too small,[1] today they charge that it's much too large to befit a divine Creator. They presume, "If God's goal was to make a habitat for humanity, he would not have made so many useless galaxies, stars, planets, comets, elements, and other components."

Physicist Victor Stenger states the skeptic's case:

> If God created the universe as a special place for humanity, he seems to have wasted an awfully large amount of space where humanity will never make an appearance.[2]

Stephen Hawking echoes this concern:

> Our Solar System is certainly a prerequisite for our existence. . . . But there does not seem to be any need for all these other galaxies.[3]

Stenger also points out that only a tiny fraction (0.0007%) of the mass of the universe is carbon. "Yet," he questions, "we are supposed to think that God specially designed the universe so it would have the ability to manufacture in stars the carbon needed for life?"[4] (see sidebar "Why So Little Carbon?"). He claims, "Energy is wasted, too. Of all the energy emitted by the sun, only

---

Why So Little Carbon?

Without carbon, physical life is impossible. No other element displays the rich chemical behavior needed to form the range of complex molecular structures life requires. Given that physical life must be carbon-based, why would God make a universe with so little carbon?

Researchers have found that the quantity of carbon must be carefully balanced between just enough and not too much because carbon, though essential for life, can also be destructive to life. Too much carbon translates into too much carbon dioxide, carbon monoxide, and methane. In large quantities, these gases are poisonous. In modest quantities, their greenhouse properties keep the planet sufficiently warm for life. In larger quantities, they can heat a planet's surface beyond what physical life can tolerate.

One of the wonders of Earth is that it is sufficiently carbon-rich *and* carbon-poor. It carries enough carbon for life but not so much as to interfere with life's atmospheric needs, such as the appropriate pressure and density for efficient operation of lungs and a temperature range (and variability) that supports a wide diversity of active, advanced species.

---

two photons in a billion are used to warm Earth, the rest radiating uselessly into space."[5]

Few people can relate to what astronomers and physicists face every day—their measurements of how vast, massive, energetic, and ancient the universe really is. Its features, including its enormity, are virtually impossible to visualize. Perhaps imagining the cosmos as a vehicle—say, a car—can help.

Like an automobile, the universe:

1. has a mass density that can be measured;
2. appears to have been manufactured to certain specifications;
3. carries passengers;
4. burns fuel and emits exhaust;
5. moves forward (though it cannot reverse);
6. is capable of slowing down and speeding up (though not of standing still); and
7. won't run forever.

This analogy is not perfect, but no illustration is. For that matter, neither is any car. I'm reminded of that fact whenever I detect a ping in my engine or notice a puddle of some colorful liquid on my garage floor. I'm reminded again as I turn into the parking lot most weekdays at Reasons to Believe (RTB).

No two cars there are exactly alike because no particular make or model meets everyone's transportation needs, preferences, and budget. Dave, Bob, Esther, Patti, Michelle, Scott, Diana, Phil, and the rest of the RTB staff chose their wheels based on the principle of optimization—what's best for each of them, all things considered. By contrast the universe is optimized for every human being—a one-year make-and-model best suited for fulfillment of *all* God's purposes.[6]

Visibility

One of the great wonders of the universe is an amazing gift that most people take for granted: the ability to see into the distance. Clarity makes an astounding difference when it comes to exploring, measuring, and understanding the cosmos. The more astronomers learn about the universe, the more they recognize how remarkable it is that *the multiple cosmic characteristics that make human life possible also make the universe visible, knowable, and measurable.*

If the universe were any smaller or larger, younger or older, brighter or darker, more or less efficient as a radiator, and if human observers were located where most stars and planets reside, the view would be so blocked as to give few (if any) clues about what lies beyond. We would be blind to the realm we live in! More importantly, no one would even be around to see it.

Additionally, the visibility and measurability of the universe from humanity's specific time window and from Earth's specific location are, by themselves, indicators of a Creator with a purpose for both the cosmos and humanity.[7]

Super-Sized

Fortunately, humans are present to take advantage of a rare moment in cosmic history and an ideal location in cosmic geography from which to gaze out over the vast expanse of the universe. As a result, scientists can uncover its secrets. But why would the universe need to be so big if just one relatively tiny planet with its population of humans is the focal point of God's creation? Why all the rest of this stuff? Exactly how much of it can be accounted for?

Until recently astronomers had no accurate measure of the number of stars and galaxies in the observable universe (see "Difference between the Observable and the Actual Universe"). But this situation changed in 2005, when scientists

---

Difference between the Observable and the Actual Universe

The universe that exists today is different from the universe that astronomers actually observe. Astronomers look *back* in time when they look at distant objects because light (even though it moves very fast) takes time to travel through space. Thus, the universe astronomers observe is the universe of the past. The farther away astronomers look, the farther back in time they see. So, for example, when astronomers produce an image of a cluster of galaxies 2 billion light-years away, that image shows them what the cluster of galaxies was like 2 billion years ago.

In a continuously expanding universe, the universe of the past is spatially smaller than the universe of the present. And more stars form as it continues to expand. Therefore, the observable universe is spatially smaller and contains fewer stars than the actual universe. (Note that all the matter and energy of the universe, including the stars and galaxies, are confined to the surface of the universe.) How much smaller depends on the geometry of the space-time surface of the universe. While the geometry of the universe isn't yet known with the degree of accuracy astronomers hope to acquire, astronomers do know that the actual universe of the present must be at least an order of magnitude (a factor of ten) larger than the universe they observe via telescopes.

---

aimed the Hubble Space Telescope at a little patch of sky no bigger than one-tenth of the Moon's diameter. Astronomers collected light from this region for a million seconds (278 hours), the longest exposure ever taken by any telescopic "camera." With an exceptionally deep look into the heavens, labeled the Hubble Ultra Deep Field, astronomers successfully imaged all the galaxies (or at least all the larger-than-dwarf galaxies) that could possibly reside in the region—including the very first galaxies that formed in the universe.[8]

The Hubble Ultra Deep Field (see figure 4.1) showed astronomers slightly more than 10,000 galaxies. Given the observed uniformity of the cosmos on large-distance scales (which is a fundamental requirement for life in the universe[9]), researchers could then do the math: more than 10,000 galaxies multiplied over the whole area of the sky totaled 200 billion galaxies in the

Figure 4.1. The Hubble Ultra Deep Field
This image produced by the Hubble Space Telescope is the deepest cosmic penetration ever achieved by an optical telescope. Every spot or smudge depicts a galaxy—except for the few foreground stars identifiable by cross-like optical defects. Some of these galaxies are over 13 billion light-years away. At this distance, they must be among the first galaxies ever to form in the history of the universe. *Image credit: NASA and the Hubble Space Telescope Institute*

observable universe. These 200 billion galaxies contain, on average, about 200 billion stars each. So, the total number of stars in these galaxies adds up to about 40 billion trillion—and that's without the estimated 10 billion trillion stars contained in the unobserved dwarf galaxies. Somewhere around 50 billion trillion stars make their home in the observable universe.

That's a mind-boggling number. A comparison may make it more comprehensible: if that same number of dimes were packed together as densely as possible and piled 1,500 feet high (as high as some of the world's tallest skyscrapers), they would cover the entire North American continent.

Here's another attempt at comparison. Shrink an average star (about a million miles in diameter) down to the size of a grapefruit. Hold that grapefruit and ask a friend to hold another. Given the average distance between stars in the Milky Way Galaxy (about 40 trillion miles), can you guess where your friend would have to take her grapefruit to illustrate the distance between stars? If you, with grapefruit in hand, stood in downtown Los Angeles, she would have to travel to Peru or Siberia. Now try to imagine that distance multiplied 40 million times (that's the necessary diameter of a volume large enough to accommodate 50 billion trillion grapefruits). Given all the empty space between galaxies, the universe on this stars-as-grapefruits scale would actually be much bigger than this calculated diameter!

The Matter of Mass

Volume gives one indication of the universe's enormity. Mass density gives another. For example, the cars in RTB's parking lot can be sized up by their dimensions (volume) *or* by their weight (more accurately referred to as mass). Two of the biggest cars in our lot are Fuz's Ford station wagon and Phil's Mercedes sedan (obviously, he worked at a for-profit company before joining the ministry staff). Fuz's car is "vast" in terms of length, width, and height. Phil's car, on the other hand, is smaller in size but more "vast" in terms of mass density (weight per volume, roughly). In a collision with a semi, Phil's car would likely fare better than Fuz's because it weighs more.

The universe is much larger than its observable volume, more vast than its observable number of stars and galaxies would indicate (see "Difference between the Observable and the Actual Universe"). What's more, the totality of its stars—those seen, those unseen, and those long-ago burned out—account for just 1 percent of the universe's total mass.[10] No wonder some people consider this incredible enormity a "waste" if humans are the main focus of the universe's existence. However, if humans are to exist, this enormous mass is critical.

Right Mass, Right Elements

Anyone who hasn't had the privilege of studying astrophysics may not realize that the universe *must* be as massive as it is or human life would not be possible—for at least two reasons. The first concerns the production of life-essential elements.

The density of protons and neutrons in the universe relates to the cosmic mass, or mass density. That density determines how much hydrogen, the lightest of the elements, fuses into heavier elements during the first few minutes of cosmic existence. And the amount of heavier elements determines how much additional heavy-element production occurs later in the nuclear furnaces of stars.

If the density of protons and neutrons were significantly lower (than enough to convert about 1 percent of the universe's mass into stars), then nuclear fusion would proceed less efficiently. As a result, the cosmos would never be capable of generating elements heavier than helium—elements like carbon, nitrogen, oxygen, phosphorus, sodium, and potassium, which are essential for any kind of physical life. On the other hand, if the density of protons and neutrons were slightly higher (enough to convert significantly more than 1 percent of the mass of the universe into stars), nuclear fusion would be too productive. All the hydrogen in the universe would rapidly fuse into elements as heavy as, or heavier than, iron. Again, life-essential elements (carbon, nitrogen, oxygen, etc.), including hydrogen, would not exist.

Right Mass, Right Expansion Rate

The second reason the universe must be hugely massive concerns its expansion rate. The rate at which the universe expands throughout cosmic history critically depends on its mass density. According to the law of gravity, the closer any two massive bodies are to one another, the more powerfully those bodies attract each other. Therefore, the closer various bits and pieces of mass are to one another in the universe, the more effectively they will slow down the universe's expansion. Conversely, the farther apart those bits and pieces are, the less "braking effect" gravity has on cosmic expansion.

Without any additional cosmic density factors such as dark energy (see "Dark Energy" later in this chapter), a universe with less mass density would not form stars like the Sun and planets like Earth. Its expansion would be so rapid that gravity would not have the opportunity to pull together the gas and dust to make such bodies. Yet if the cosmic mass density were any greater, gas and dust would condense so effectively under gravity's influence that all stars would be much larger than the Sun. Any planets such stars might hold in their orbit would be unsuitable for life because of the intensity of the stars' radiation and because of rapid changes in the stars' temperature, radiation, and luminosity—not to mention the radiation and gravitational disturbances caused by neighboring supergiant stars.

With only a little extra mass, the universe would expand so slowly that all stars in the cosmos would rapidly become black holes and neutron stars. The density near the surface of such bodies would exceed five billion tons per teaspoon (1 billion tons per cubic centimeter). At such enormous densities, molecules are impossible. So are atoms. Therefore, life would be impossible.

The radiation and gravitational disturbances from such black holes and neutron stars would also make physical life impossible anywhere in such a dense universe. Physical life cannot exist in a universe with a mass density any less or any more than the actual cosmic value.

Some might argue that a sheer coincidence explains the particular mass density of the universe and, therefore, that the universe's mass implies nothing about intentionality. However, the mass of the universe is fine-tuned to provide two life-essential features simultaneously: (1) the just-right amounts and diversity of elements, and (2) the just-right expansion rates throughout cosmic history so that certain types of stars and planets form at the just-right times and in the just-right locations. Fine-tuning to provide two life-essential characteristics at once hints louder than a whisper at purposeful design. So, too, does the high degree of fine-tuning.

An Exquisite Balance
While stars and planets account for only about 1 percent of the total matter (hence mass) of the universe, even that small percentage must be extraordinarily fine-tuned for life to exist. Picture a huge vehicle—something much bigger than a car. Maybe the US Navy's aircraft carrier the USS *John C. Stennis* (see figure 4.2). Now imagine a tiny fleck of paint from that ship, so small against your hand you can barely see it. If such a vehicle were compared to the universe in its earliest moments, removing that speck or adding an extra drop of paint to it would be enough to alter the vehicle's mass so much as to make it completely useless for transporting passengers.

In reality, the delicacy of that ratio is far more extreme than the ship analogy reveals. For the reasons noted previously, and *if* no other density factors influence the expansion of the universe, at certain early epochs in cosmic history, its mass density must have been as finely tuned as one part in 10^{60} to allow for the possible existence of physical life at any time or place within the entirety of the universe.[11] This degree of fine-tuning is so great that it's as if right after the universe's beginning, someone could have destroyed the possibility of life within it by subtracting a single dime's mass from the whole of the observable universe or adding a single dime's mass to it.[12]

Figure 4.2. The USS *John C. Stennis*
This United States Navy aircraft carrier is 1,092 feet long. A self-contained city, it has a displacement of 100,000 tons when fully loaded. If the USS *John C. Stennis* were as fine-tuned as the universe, adding or subtracting a billionth of a trillionth of the mass of an electron from the total mass of the aircraft carrier would make it uninhabitable. *Image credit: United States Navy. Photo taken by US Navy Mass Communication Specialist 3rd Class Paul J. Perkins.*

Recently, astronomers have discovered evidence that other cosmic density factors *do*, in fact, influence cosmic expansion (see table 4.1). These factors, while reducing the degree of fine-tuning in the cosmic mass density, introduce far more spectacular fine-tuning elsewhere.[13] The illustration of adding or subtracting a dime's mass to the universe is too conservative. The fine-tuning of the cosmic density parameters is far more impressive.[14]

Dark Matter
Astronomers now recognize that, in addition to a specific cosmic mass with quantities of protons and neutrons precisely fixed to make life possible, every component that makes up the universe, both matter and nonmatter, must be present at a specified value or physical life would not exist.

Table 4.1 Inventory of All the Stuff That Makes Up the Universe	
Cosmic Component	Percentage of total cosmic density
Dark energy (self-stretching property of the cosmic space surface)	72.1
Exotic dark matter (particles that weakly interact with ordinary matter particles and light)	23.2
Ordinary dark matter (particles that strongly interact with light)	4.35
Ordinary bright matter (stars and star remnants)	0.27
Planets (a subset of ordinary dark matter)	0.0001

Note: This inventory began with an exhaustive compilation by Princeton cosmologists Masataka Fukugita and James Peebles. It was based initially on the best measurements prior to 2005.[15] Updates were made possible in 2006 and 2008 by the second and third releases of the Wilkinson Microwave Anisotropy Probe's (WMAP) results.[16]

Little more than half a century ago, astronomers came to realize that the stuff they see through their telescopes makes up only a tiny fraction of the total amount of matter in the universe.[17] As that realization dawned, astronomers hypothesized that this "dark matter" was made up of cold gas and failed stars[18] ("brown dwarfs," stars with so little mass they never ignite nuclear fusion[19]). When it became apparent that the maximum contribution from cold gas and failed stars was grossly inadequate to account for the total, Princeton University astronomer James Peebles proposed that the dark matter was composed of small rocks. Peebles's proposal led to a familiar joke among astronomers that "the universe either was Peebled with pebbles or pebbled with Peebles." Ongoing research has led to the development of an inventory, a list of the various components that comprise the universe—both matter and energy (see table 4.1).

After acknowledging that the total quantity of protons and neutrons in the universe must be set at a precise value for the universe to produce the right kinds and quantities of life-essential elements at the right times in cosmic history, astronomers discovered yet another fine-tuned feature. The proportion of ordinary bright matter (the protons and neutrons that form stars) with relation to ordinary dark matter (the protons and neutrons that form gas, dust, rocks, and planets) must also be fine-tuned for life. Too much or too little of the bright stuff would expose potential life-forms to either too much or too little light, heat, and radiation, for example. The abundance of life-essential elements and of radioactive isotopes is also affected by this balance. As it turns out, life requires that 5 percent of ordinary matter be bright and 95 percent dark.

But ordinary matter does not add up to enough mass to generate the required expansion history of the universe. This history must be just right, with the universe expanding at the just-right rates at the just-right times, for life to be possible. Some other form of matter must have been available in a precise abundance and location. Theorists calculated that this other matter, now called exotic matter (comprised of particles that interact only weakly, if at all, with ordinary matter particles and light), would need to be nearly five times more abundant than the ordinary matter, for life's sake. And that's what researchers have found.

In other words, not only must the universe be as massive and vast as it is (given the cosmic mass density and expansion time required to make a planet like Earth), but also each of the different mass components must be neither smaller nor larger than they are.[20] However, not all of what makes up the universe is matter.

Dark Energy
Much in the same manner as a car, the universe has separate systems for slowing down and speeding up. Gravity functions as the main braking system. During the early era of the universe, before its mass became widely dispersed, gravity effectively applied the brakes on the universe, slowing down its expansion from its initial creative burst. This creation event is familiarly referred to as the big bang.

A bizarre feature called "dark energy" (discovered so recently that the scientific community still hasn't settled on exactly what to call it) serves as the acceleration system. Perhaps this quality is best described as a self-stretching property of the cosmic surface (the spatial surface of the universe along which all matter and energy are distributed). For the first approximately 7 billion years of its existence, the universe expanded at a decelerating rate. Then, as the components of the universe gradually spread apart, gravity became progressively weaker in its capacity to slow down the expansion, and dark energy gradually became stronger or more effective in its capacity to accelerate the expansion. This dark energy effect may be described as a self-stretching capacity that increases as the surface of the universe continues to expand (see "What Is Dark Energy?").

Eventually dark energy's accelerating effect overtook the braking effect of gravity, and for approximately the last 6.7 billion years, cosmic expansion has been speeding up.[21] In the future, the effect of gravity will become increasingly weaker in its capacity to slow down cosmic expansion while the effect of dark energy (assuming no radical alteration in its future behavior) will grow increasingly stronger. As a result, the universe will continue to expand at an accelerating rate.

The observational verifications that dark energy is the predominant component of the universe and, therefore, that the universe will expand at an ever-increasing rate put an effectual end to the oscillating universe model and to the Hindu/Buddhist concept of a reincarnating universe.[22] Accelerating cosmic expansion means the universe can never contract; therefore, it cannot rebound. This fact eliminates the possibility of a renewal, rebirth, or second beginning for the universe.[23]

Much in the same way as the mass density of the universe must be fine-tuned to ensure that the universe expands in exactly the way life requires, the dark energy density also must be exquisitely fine-tuned. However, its fine-tuning is orders of magnitude more stringent. If dark energy were changed by as little as one part in 10^{120}, the universe would be unable to support life.[24] A number that small can be hard to picture. If compared to the mass of the entire universe, it would be no bigger than a billionth of a trillionth of a trillionth of an electron's mass.[25]

Clear and Present Purposes

Both cosmic mass density and dark energy density hugely impact not only the possibility for human life but also the possibility for individuals to observe, explore, and understand the universe. Given the particular laws and constants of physics that govern the universe, the possibility for life and discovery mandate

What Is Dark Energy?

Though often described in popular literature as an antigravity force, dark energy is not a force. A better, though still imperfect, analogy would be to describe it as the opposite of the effect you feel when stretching an elastic band.

The more an elastic band is stretched, the more energy it gains to encourage its contraction. Thus, the more someone stretches the band with his fingers, the more he feels the tension that impels the band to contract.

The surface of the universe acts the opposite way—it is like a gigantic elastic band that *wants* to expand outward. The more the cosmic surface stretches, the more energy the surface gains to propel even more stretching of the surface. When the universe was very young, gravity kept it from expanding much more rapidly from its initial infinitesimally small volume. At that time, dark energy would have been relatively weak in its capacity to expand or stretch out the space surface of the universe.

However, as the universe grew older and older, the cosmic space surface became bigger and bigger. This increasing size of the cosmic surface meant much more dark energy became available to expand the cosmic space surface.

Thus, the entirety of the universe eventually began to expand at an accelerating rate. That's because all the universe's matter and energy, as well as its space-time dimensions, are confined to its surface.

Astronomers have yet to determine the nature of dark energy precisely enough to make a confident pronouncement. It is even possible, though not probable, that two or three factors contribute to the dark energy effect. The latest studies favor a single factor. They reveal that the dark energy density is roughly constant throughout cosmic history, or at least all but its earliest moments.[26]

that the universe be vast in all ways, including volume and mass, at the particular epoch during which intelligent life exists.

The skeptic might say it's conceivable, though not necessarily possible, that a different set of laws, constants, and dimensions might provide humanity with an acceptable habitat without requiring such an incredibly vast universe. There is, however, the perspective of purposes beyond mere provision of a habitable environment. These additional purposes are reflected in the specific set of physical laws, constants, and dimensions the universe manifests. Given this unique set, the universe must indeed be as vast and as massive as it is.

And though its enormity strains the human capability to imagine, that vastness says something about the high value of and high purposes for humanity's existence. Rather than seeing ourselves as insignificant specks in the immensity of the cosmos, we can consider that immensity an indicator of our worth. It seems the Creator invested a great deal—a universe of 50 billion trillion stars, plus a hundred times more matter, all fine-tuned to mind-boggling precision—for us. If not for the strength and abundance of evidence in support of that notion, it would seem the height of arrogance. Humility demands that we take a deeper and wider look at that evidence.

Chapter 5

The Origin and Design of the Universe
by Jeff Zweerink

Prominent atheist cosmologist Lawrence Krauss is fond of making statements like this:

> The amazing thing is that every atom in your body came from a star that exploded. And, the atoms in your left hand probably came from a different star than your right hand. It really is the most poetic thing I know about physics: You are all stardust. You couldn't be here if stars hadn't exploded, because the elements—the carbon, nitrogen, oxygen, iron, all the things that matter for evolution—weren't created at the beginning of time. They were created in the nuclear furnaces of stars, and the only way they could get into your body is if those stars were kind enough to explode. So, forget Jesus. The stars died so you could be here today.[1]

Obviously, Krauss's statement is specifically antagonizing to Christians, but I think it is offensive to *all* religions that believe in a God who created the universe and all it contains. Krauss argues that science can explain everything; therefore, we don't need a god.

However, as a Christian who is a scientist, I come to a different conclusion. A theistic worldview provides the best explanation of our scientific understanding of the universe. Let me say that again: a *theistic* worldview provides the best explanation of our scientific understanding of the universe. The evidence for this conclusion is plentiful, but here I'll share just three powerful examples.

In the Beginning
At the start of the 1900s, three significant features characterized scientists' understanding of the universe. First, the universe was eternal and had existed forever. Second, the universe was static and unchanging on the largest scales. And third, as one moved through the universe, the laws of physics changed in subtle ways. Before describing how twentieth-century scientific advances changed this picture, I want to contrast this early scientific view with the Bible's description.

Starting in Genesis 1 we see that God created the heavens and the earth. In this account, God brought the universe into existence out of nothing. In other words, the universe had a beginning; it has not existed forever. Other prophets also speak about the nature of the universe. For example, Isaiah records God describing himself as the "Maker of all things, who stretches out the heavens ... by myself" (Isaiah 44:24). Not only does this text affirm a beginning for the universe, it also implies that the universe is dynamic on the largest scales rather than static and unchanging.

What might be most important scientifically is a statement the prophet Jeremiah recorded. According to Jeremiah 33:25, God declares that "if I have not made my covenant with day and night and established the laws of heaven and earth," then he would break his promises and not fulfill what he said he would do. Although this passage refers to how objects move through the heavens, we now know that this motion ultimately depends on constant laws of physics! So, Jeremiah is describing a universe where the laws of physics are constant, which is a critical criterion for the development of the scientific enterprise. Note the contrast between the early twentieth century's scientific picture of the universe and the one revealed by God. A little more than 100 years ago, scientists thought the universe was eternal, static, and governed by changing laws of physics. God revealed that the universe is temporal, dynamic, and governed by constant laws of physics.

So, let's look at some of the important discoveries that brought the scientific picture in line with the theistic one described by the Bible. During the 1910s, Albert Einstein recognized that in the scientific description of the day, the laws of physics changed as one moved through the universe. Philosophically, he didn't like that idea. So, he set about developing a model of the universe where the laws of physics were constant. In doing so, he developed the special theory of relativity and the general theory of relativity. The key feature of these theories is that the laws of physics are constant and unchanging regardless of how you're moving or where you're located in the universe. Scientists have thrown numerous experimental tests at the theory of general relativity to see if it is valid or not, and it has passed every test with outstanding success. It is one of the best-established and best-accepted scientific theories known today.

One consequence of general relativity, when you solve its equations, is that the universe ought to be dynamic, either contracting or expanding. Initially, Einstein didn't like this idea, but measurements in the 1920s and 1930s ultimately established that the universe is indeed expanding. Edwin Hubble looked out at what he called island universes (we now call them galaxies) and found that the farther away a galaxy was, the faster it was moving away from us. This is a telltale signature of an expanding universe. General relativity predicted an expanding or dynamic universe, and measurements of these distant galaxies show that the universe *is* expanding. If it's expanding, then perhaps if you run time backward there was a beginning.

Scientists resisted this idea for quite some time, and they looked for numerous ways to retain the concept of an eternal universe. But in the 1960s, with the measurement of the cosmic microwave background radiation, Stephen Hawking, Roger Penrose, and other scientists developed some very powerful theorems. These theorems showed that if general relativity describes the dynamics of the universe accurately (and it has passed every test we have thrown at it) and if the universe contains mass (and we're pretty sure of that), then we can draw the conclusion that when you run time backward the universe has a boundary. In other words, the universe began to exist.

So, at the close of the twentieth century, the scientific view of the universe looked very much like a theistic worldview thanks to significant scientific discoveries. *We live in a universe that began to exist. The universe is expanding. And constant laws of physics govern the universe.* These are the three essential features of all big bang models. In other words, the universe that God revealed to us through the Bible matches the universe that we see when we study creation.

In recent years, scientists have proposed multiverse models—where our universe is just one of a great ensemble of universes. Some people argue that the multiverse challenges the notion of the beginning. However, even the existence of an inflationary multiverse affirms the conclusion that the universe began to exist.

The first piece of scientific evidence supporting a theistic worldview is that inflationary big bang cosmology shows that the universe began to exist. This conclusion supports the second premise of the kalam cosmological argument that, in syllogism form, says:

Whatever begins to exist has a cause;
the universe began to exist;
therefore, the universe has a cause.

Fine-Tuned Coincidences

The second major support for the existence of a Creator is the pervasive evidence of design in the universe. Such evidence appears throughout the universe, from the laws of physics to the genetic code.

Consider what it takes for humanity to live. For starters, we (and indeed all organisms) require carbon, water, and a planet where liquid water can exist in the presence of abundant carbon. Many of the scientists studying the universe's ability to support life have concluded that the universe looks like it was designed to do so. Even self-professed atheists and agnostics have acknowledged the appearance of design. Atheist astronomer Fred Hoyle said, "A common sense interpretation of the facts suggests that a superintellect has monkeyed with physics, as well as with chemistry and biology. . . . The numbers one calculates from the facts seem to me so overwhelming as to put this conclusion almost beyond question."[2]

It is no controversy to say that the best scientific evidence indicates that the universe appears designed for life, and we see that evidence across the scientific disciplines. Let's look at some of the universe's features that look designed to support life.

Fine-Tuned Dimensions

We live in a universe with three large spatial dimensions and one time dimension, but we can analyze what would happen if the dimensions were different. For example, what if there were only *two* spatial dimensions? As it turns out, if there were two or fewer spatial dimensions, then the universe would not be complicated enough for life. Imagine an animal in two dimensions. If the animal has a passage for food intake and a different passage for expelling waste, in two dimensions such passages would cut the animal in half. Now you could make an argument that the food could come back out the same way, but such an argument misses the point. In two dimensions there are not enough ways physically to make connections in order to have the complexity life requires.

Perhaps more dimensions would be better. It turns out that if you go to four, five, or more spatial dimensions, there are no stable orbits. With more spatial dimensions, atoms are not stable and planets either spiral into their host star or fly into space. Thus, the critical requirements for life—elements and planets—don't exist. Changing the number of time dimensions creates even worse problems. Anything more or less than one time dimension leads to a situation where physics is unpredictable. Now you may say, "I don't know how to do physics, so why is it necessary for it to be predictable?" It so happens that "physics being predictable" is a foundational requirement for life. Every organism must be able to sense the environment, tell what has occurred in the past, and determine what will happen in the future. However, if physics is *un*predictable, this is impossible. Life can exist only in a universe with three large spatial dimensions and one time dimension.

Fine-Tuned Physics
Let's turn our attention now to the laws of physics, especially at how carbon, oxygen, and hydrogen exist in the universe. According to big bang cosmology, in the first four minutes of the universe's beginning, only hydrogen and helium (and trace amounts of lithium and beryllium) existed. All the elements heavier than these are formed inside stars. As scientists studied how well stars produce carbon and oxygen, they found huge deficiencies of these elements unless some remarkable coincidences were true. In particular, the difficulty of producing carbon is that three helium nuclei have to come together at the same time. Because it's a three-body interaction, that makes the process very slow. However, as scientists looked more closely, they recognized two important factors that allow the formation of carbon.

First, when two helium atoms come together, they can form a beryllium-8 nucleus. Beryllium-8 is not stable, but it does stick around for a while. This means that only one more helium nucleus needs to collide with the beryllium-8 to form carbon, which speeds up the reaction considerably. However, even with the metastable beryllium-8 nucleus, stars would not produce enough carbon. Something else was missing. In the 1950s, Fred Hoyle, one scientist working on the problem, recognized a solution that would produce carbon rapidly enough. If carbon had a particular energy level just above the ground state (or the lowest energy state), then the reaction would proceed more quickly. At the time, carbon was not known to have this particular energy level, but subsequent study by Hoyle and others showed that such an energy level indeed existed. Without a stable beryllium-8 nucleus and a finely tuned energy level for carbon, the universe would not produce sufficient carbon for life.

Yet, for the universe to contain sufficient carbon, oxygen's energy level must be fine-tuned, too. If oxygen had an analogous energy level, then all carbon would fuse into oxygen, leaving no carbon. Fortunately, oxygen does not have this energy level. These coincidences show that the universe contains large amounts of carbon *and* large amounts of oxygen, both of which are critical for life.

Although I've used the word "coincidence" to describe these essential features, I think it is more accurate to say that the universe was *designed* to produce the elements necessary for life. It was the discovery of the exquisite balance of the laws of physics that prompted Hoyle to acknowledge the appearance of design.

Similar design features also ensure the universe has the necessary hydrogen for life. For example, if two protons could join and form a stable nucleus, the universe would have used up all the hydrogen in the first few minutes after its creation. The same thing would have happened if there were any stable five-nucleon elements or eight-nucleon elements. The strength of the fundamental laws of physics determines whether these design features exist or not. If we plot the strength of the electromagnetic interaction against the strength of the strong nuclear interaction, then we can ask the question: where in this space can life exist? The answer is that only a very small region of the available parameter space meets all the necessary requirements for life. This feature points to design, that there is a Creator who fashioned the universe for a purpose.

Fine-Tuned Moon
Looking a little closer to home, we see evidence of design in the Moon that orbits Earth. Jupiter and Saturn both have satellites that are larger than Earth's Moon. However, when compared to the size of the host planet, Earth's Moon is in a class by itself. The Moon's large size plays an important role in Earth's capacity to host life. For example, without such a large moon to maintain stability, Earth's rotation axis would wobble, causing violent and catastrophic changes to our climate. Perhaps more importantly, the size of the Moon helps provide the heat necessary to enable plate tectonic activity on Earth for billions of years. The gravitational tug of the Sun and the Moon causes Earth's interior to flex, stretch, and compress. This tugging heats up Earth's interior, and that heat drives the plate tectonic activity. As scientists seek to understand how Earth acquired such a large moon, they recognize that it took a remarkable collision early in Earth's history. This collision needed to happen at the just-right speed, at the just-right time, at the just-right angle, with an object of the just-right size. It really is an unusual collision. It looks like the Moon was designed so that Earth can support life.

Fine-Tuned Genes

A third piece of evidence for fine-tuning comes from biology. DNA is commonly called the blueprint of life. It is made up of four compounds represented by the letters T, C, A, and G. These letters come in groups of three and each group specifies the production of an amino acid. Sequences of three letters where each letter has four options mean there are 64 different possibilities. Yet, because only 20 different amino acids are involved in life, this means that different combinations of three will produce the same amino acid. Likewise, the amino acid sequences determine how proteins will fold, and sometimes different amino acids will still produce the same protein folding.

Scientists can then ask the question: how well does the genetic code ensure that proteins fold and function properly even with mutations in specific letters? Given the redundancy in amino acid coding and the similarity of amino acids in protein folding, scientists have determined that our genetic code is literally one in a million in its ability to minimize errors caused by mutations. Additionally, the genetic code resides at a global optimum (not just the best in a given range, but the best overall). Such exquisite fine-tuning reminds me of a statement by Francis Crick, codiscoverer of the iconic DNA double helix: "Biologists must constantly keep in mind that what they see was not designed, but rather evolved."[3]

I disagree. When scientists look at the universe, they see evidence of fine-tuning and design at all scales. They see fine-tuning in the fabric of space-time, in the form and strengths of the laws of physics, in the size of our Moon, in the genetic code, and in many other areas I have not mentioned. It seems natural to conclude that when we see design it is because a designer exists and created the universe to support humanity.

The Philosophical Foundations of Science

Inflationary big bang cosmology shows that the universe began to exist. Science also shows that the universe looks like it was designed for life to exist. Now let's look at the third piece of evidence pointing to God's existence: the foundations of the scientific enterprise.

No one disputes the utility of science. We have discovered great things about this universe, and we have developed great resources and technology to benefit humanity. For science to work, though, every scientist must operate under a specific set of philosophical assumptions. For example, a scientist must believe that the laws of nature are uniform throughout the physical universe. If the laws were not uniform, then no measurements made here on Earth would apply to the universe as a whole. Without uniform physical laws, the scientific enterprise would have never even gotten started.

A scientist must also assume that the physical universe is a distinct and objective reality. Contrast this view with Hinduism, for example, where the physical world is ultimately an illusion. If the world were an illusion, why would we go about studying it and trying to understand how it works?

Additionally, a scientist must assume that the laws of nature exhibit order, patterns, and regularity. I am reminded of my studies of Greek mythology in school. Greek mythology is littered with gods and goddesses who are arbitrary and capricious. One day they're bestowing gifts and favors, the next they're angrily stirring up oceans or throwing lightning bolts. If this were the worldview a scientist has about the universe, then why would he or she expect to see any sort of order, regularity, and pattern in nature?

For science to work, the physical universe must also be intelligible. Those patterns and order must be sensible and understandable in some way. Also, the world itself must not be divine; otherwise, the proper response would be to worship it, not treat it as an object of rational study. Again, I think of the Eastern mystical religions or the "basic" religions as they're called. In these worldviews, the physical is divine and mystical, and our goal is to become one with nature, not to study it and figure out how it works.

Another assumption the scientist makes is that the world is good, valuable, and worthy of study. In college, I read a story about Siddhartha Gautama, the Buddha. During the course of the story, Siddhartha ultimately gained genuine enlightenment—perfection, the sought-after goal—through detachment from the world. He learned that the world is not good, but rather it gets in the way of the true goal. If detachment from the world is the ultimate goal, why would we try to understand the way the world works?

Productive, successful scientists must also assume that the universe could be different. In other words, if the Creator has free agency, then he could have chosen to do things differently. That's why we must study and experiment to figure out what was done. A theistic view of the world not only encourages the scientific enterprise but also compels humanity to study creation. In Genesis, God issues an imperative to Adam and Eve to take dominion over nature. Moreover, God's moral law grounds and encourages essential intellectual virtues such as honesty, integrity, a good work ethic, and many others.

One final assumption a scientist must make to do good science is that humans possess an ability to discover the universe's intelligibility. If humans did not have this ability, then it would make no difference whether the universe met all the assumptions listed previously or not. The scientific e̶ never have gotten started. Consider what atheistic naturalism say manity's ability to reason and understand. If atheistic naturalism is t̶ humanity's primary purpose is simply to survive. No reason exists to that human thought could comprehend how the universe works. C. S. Le̶ describes the atheist's conundrum very well:

> If the solar system was brought about by an accidental colli-
> sion, then the appearance of organic life on this planet was also
> an accident, and the whole evolution of Man was an accident
> too. If so, then all our present thoughts are mere accidents—
> the accidental by-product of the movement of atoms. And this
> holds for the thoughts of the materialists and astronomers as
> well as for anyone else's. But if their thoughts—i.e., of ma-
> terialism and astronomy—are merely accidental by-products,
> why should we believe them to be true? I see no reason for
> believing that one accident should be able to give me a correct
> account of all the other accidents.[4]

Given these foundational principles of science, one should then ask the question: what worldview(s) properly anchors all these philosophical pre-suppositions? Atheistic naturalism cannot. Buddhism and Hinduism cannot. Eastern religions and Greek mythology cannot. But a theistic worldview—where God creates humanity with a purpose, moral code, and desire to worship and know God—can and does. I am *not* saying that a scientist must be a theist to do science, but I *am* saying that a scientist must adopt the worldview of a theist for the scientific enterprise to progress consistently and endure over time.

Many vocal researchers claim that science falsifies the notion of a God and justifies an atheistic belief. However, a close inspection of the latest scientific evidence provides strong evidence that God exists. Throughout the Bible, multiple authors describe a universe that began to exist, is governed by constant laws of physics, and that expanded over time. Inflationary big bang cosmology affirms that we live in a universe matching this description—*even if* our universe is part of a larger multiverse. Our universe shows exquisite fine-tuning for life to exist. It looks designed for life! Only a theistic worldview anchors all the essential philosophical foundations required for science. Not only does the scientific evidence demonstrate the rationality of belief in God, I would argue that God's existence provides the best explanation of our scientific understanding of the universe.

nterprise would
s about hu-
rue, then
think
vis

Chapter 6

Would the Discovery of ET Disprove Christianity and Must ET Exist?

by Jeff Zweerink

According to popular narrative, Copernicus's assertion that Earth orbited the Sun started a relentless process of scientific discovery serving to remove any thought that humanity occupied a special place in the universe. Earth was not the center of the solar system, the Sun was not the center of the galaxy, and the Milky Way Galaxy was not the center of the universe. Many people take this idea even further to claim that Earth, including the life inhabiting it, is completely mediocre in every way. The discovery of any kind of life—especially intelligent life—beyond the confines of Earth would forever falsify the notion that humanity has any claim of exceptionalism.

Associated with this line of reasoning, many believe that discovering extraterrestrial (ET) life would powerfully demonstrate that many of the world's religions, particularly Christianity, cannot be true. After all, the Bible mentions nothing about life beyond Earth! So, would the discovery of ET disprove Christianity? It's possible, but not likely.

Essentials of Christianity

Before evaluating Christianity's effect on finding ET, one must understand the foundation of the Christian faith. Unless finding ET undermines the basis of Christianity or directly contradicts a teaching of the Bible, the hypothetical discovery cannot disprove the faith. So, what are the basics of the Christian faith?

Creation, Fall, Incarnation, Resurrection, and New Creation

Throughout history, Christians have sought to articulate the essential beliefs of the faith. The carefully chosen wording of the Apostles' Creed summarizes them in this way:

I believe in God, the Father almighty, *creator of heaven and earth*. I believe in Jesus Christ, God's only Son, our Lord, who was *conceived by the Holy Spirit, born of the Virgin Mary*, suffered under Pontius Pilate, was crucified, died, and was buried; he descended to the dead. On the third day *he rose again*; he ascended into heaven, he is seated at the right hand of the Father, and he will come to judge the living and the dead. I believe in the Holy Spirit, the holy catholic Church, the communion of saints, the forgiveness of sins, *the resurrection of the body, and the life everlasting*. Amen. (emphasis added)[1]

Creation. Genesis 1:1 declares "In the beginning God created the heavens and the earth." The Hebrew word for created, *bārā'*, carries the connotation that God brought the universe into existence out of nothing (creation *ex nihilo*). In other words, God is the maker of the universe. He determines how the universe operates, decides how to reveal himself to humanity, provides the standard of right and wrong, and judges the standing of each human. One primary function of the Bible is to communicate God's revelation of himself to humanity, particularly how we relate to him. Because we are the creation and God is the Creator, our role is to submit our actions to his direction.

The Fall. After God created the universe and fashioned Earth as a habitat for humanity, he created the first humans, Adam and Eve, and placed them in the Garden of Eden. One of the couple's first recorded acts was to disobey God's command and eat of the forbidden tree. God is just; hence, their rebellion required judgment. Adam and Eve served as humanity's representatives in the Garden and rejected God's command, resulting in humanity's eternal separation from God.

The Incarnation. Though God is just, he is also compassionate. The only option for humanity after Adam and Eve's sin was eternal separation. However, God chose to reveal himself in an incredible way. Jesus, the second person of the Trinity, came to Earth as a man. He did not cease to be God (that would violate the character of God as well as the rules of logic) but took on a human nature. In theological terms, Jesus was fully God *and* fully man in hypostatic union. As a human being without sin, he could bear the punishment for sin required by God's justice.

The Resurrection. Two millennia ago, when Jesus Christ walked the earth as a man, his life ended in a brutal crucifixion. His sinless life enabled his death on the cross to atone for the sin of humanity. However, the story does not end there. On the third day, Jesus rose from the dead, validating the biblical claim that he was God.

The New Creation. Jesus commissioned his followers to go to the ends of the earth so that everyone can know how God redeemed humanity from its rebellion. All who receive Christ's atonement by faith become part of God's kingdom. Eventually, Christ will return to Earth and usher all who belong in his kingdom into a new creation. All who refuse Christ's atoning work on the cross will spend eternity apart from God (the logical outcome of their choice), enduring life devoid of all the goodness, blessing, and fulfillment God's presence brings. Obviously, this brief description of five doctrines taken from the Apostles' Creed does not explain the entire Christian faith. Whole books are needed for that task.

One of the Bible's primary functions is to reveal this message of redemption to all humanity. Sin leaves every person on this planet condemned before a holy, righteous God. Only the sacrifice of his perfect Son can pay our debt and satisfy his justice. Faith is the gift he gives so we can be reconciled to him. Would finding life beyond the confines of Earth contradict any aspect of this redemption message?

How Would ET Affect Christianity?
The type of ET found influences the answer. Few, if any, theologians would have any concerns about finding microbial life on a distant exoplanet. From a theological perspective, the Bible makes no explicit claims about the origin of life. However, Genesis 1:2 may imply that God created life early in Earth's history and then protected it during an era when conditions on the planet were hostile for any living thing. A discovery of microbial ET would raise questions regarding life's origin and whether it requires anything beyond the normal operation of the physical laws governing our universe. It may be that God created a universe where life arises by naturalistic processes, but one can make a strong case based on our best scientific understanding that life's origin and the development of Earth's life-friendly conditions both require divine intervention.[2]

Microbial ET represents the more likely find, but discovering intelligent ET would be *far more* theologically compelling (like humanity that is "made in God's image"). The Bible is largely silent on the issue of intelligent ET (at least the physical kind), so the dominant position in historic Christian thought is that humanity represents the only intelligent physical life in the universe. Just as people have speculated on the existence of intelligent ET for centuries, theologians have also contemplated how such a finding would interact with historic Christian thought.[3] Here are some of the proposals offered, in no particular order. All these options, except for the last one, assume that God created intelligent life on other worlds and that these creatures, like humanity, chose to rebel against God.

1. Jesus's incarnation, death, and resurrection here on Earth was a singularly important happening that results in redemption for *all* intelligent ET. Perhaps humanity will spread throughout the universe, taking the gospel to all these creatures. Or maybe God reveals himself on each of these planets in a way that declares what Jesus accomplished on Earth. This option does present at least one difficulty. The Bible describes Jesus as the second Adam, meaning that both are related physically and in nature. Any intelligent ETs in this scenario have no physical relation (and maybe not even in nature) to Jesus.

2. Jesus becomes incarnate on each planet where creatures rebel, taking on the nature of the beings created by God on each planet. Although God created humanity in his image, perhaps his creations on other planets have a nature that reflects God's image in different ways. Given the description of the hypostatic union offered by Thomas Aquinas, the addition of *other* natures besides Jesus's human one seems plausible.

3. The nature of the rebellion of other intelligent ETs requires another means of redemption for them. Given that God has only revealed the redemption plan for humanity, any proposals for what these other redemption paths look like are pure speculation.

4. No redemption is possible. This option seems the most offensive to human sensibilities, specifically because of the difficulty of reconciling it with God's goodness, omniscience, and power. Some would object that an all-knowing, all-good, all-powerful God cannot create beings subject to eternal hell. In fact, many raise the same objection in the context of humanity. However, a few points warrant mention. First, our finite minds cannot comprehend all that God knows. Second, true free will has consequences. While God is good in his nature, he is also just. Although this option seems offensive, it does have precedent in the Bible. Angels had a choice to either serve God or Lucifer. No offer of redemption exists for those angels who followed Lucifer.

5. God created intelligent life with free will, but these creatures chose not to rebel as humans did. Lacking any violation of God's command, these creatures have no need of redemption. They already enjoy proper relationship with God. C. S. Lewis explores this idea in his Space Trilogy book series.

One final, and very real, possibility: God created only one intelligent creature in the entire universe. If so, then the redemption story of life on Earth is *the story* of the universe and this discussion becomes moot.

Science and Theology's Common Ground

A common objection scientists level against Christianity, and religion in general, is its apparent lack of testability. Stated another way, they charge that Christianity is so flexible and vague that nothing could ever falsify it. Ironically, many people (Christians and non-Christians alike) think that finding intelligent ET would falsify Christianity. There are, however, several ways that historic Christian theology could incorporate the discovery of intelligent ET.[4] But doesn't the wide number of options just validate the charge of inordinate flexibility and vagueness?

Actually, scientists deem that property a virtue. Consider a topic investigated extensively by the scientific community over the last hundred years: What is the proper interpretation of quantum mechanics? One might think this to be a settled question, considering that quantum mechanics is one of the two most successful scientific models ever developed (the other is general relativity). A quick perusal of the literature reveals that many different interpretations exist for the underlying nature of quantum mechanics.[5] The breadth of options on this topic represents the outstanding efforts of many scientists to address a difficult question. Developing a range of interpretations helps scientists know the experiments that will distinguish which interpretation best represents reality. Similarly, a range of models for ET's redemption helps theologians discern underlying details of God's redemptive story.

Einstein's development of his theory of relativity did not prove Newton wrong. Newtonian dynamics still describes properly the motion of the majority of objects through space. Einstein's relativity just gives a more complete picture of how gravity operates. If we find intelligent ET (and that's a *big if*), the discovery would not invalidate the historic Christian understanding of redemption. Like Einstein's relativity, the discovery of intelligent ET would give us a more complete picture of how God interacts with humanity. Jesus's life, death, and resurrection still provides the only means of redemption for fallen humans, but maybe God's redemption narrative encompasses more than just humanity.

Must ET Exist?

In a digital age we have grown accustomed to the accelerated pace of discoveries. For many people, science, engineering, and medical breakthroughs have provided and will continue to provide answers to all of humanity's challenges—thus, it will be no different here. Science will ultimately answer the question, "Is there life out there?"

But it hasn't. And for those like me who want a definitive answer, it will likely be decades before we have any data that directly addresses the issue.

An Argument in Favor of ET

Consider a photon (something like a particle of light) emitted in the early history of the universe. When the universe was just 400,000 years old, it cooled to a temperature where protons and electrons could form hydrogen atoms. Every electron that joined a proton emitted one photon of light somewhere in the visible-to-infrared range of the electromagnetic spectrum. As this photon traveled toward the location where Earth would eventually orbit, its odyssey was marked by nearly unfathomable distances and wonders.

About 200 million years later, the photon encountered the first stars to form in the universe. Scientists have not yet found the light emitted by these stars, but calculations show that they were likely giants by today's standards. These first stars had masses tens and even hundreds of times larger than our Sun. Consequently, some of them burned through their nuclear fuel in just a few million years before exploding in spectacular supernovae. The energy emitted by these stars kicked all the electrons away from their protons, thus reionizing the universe.

As the photon continued its journey, it passed the first galaxies. Often these galaxies hosted an incredibly large black hole at their centers. The largest black holes measured to date weigh in at a staggering 30–40 billion times the mass of the Sun. For comparison, the Milky Way Galaxy (MWG) hosts a black hole with a mere 4.3 million solar masses. During the early history of the universe, many of these massive black holes consumed large amounts of dust and stars and emitted copious amounts of X-rays and gamma rays.

Over the next 10–12 billion years, the photon passed galaxy clusters so large that it would take over 500 million years to traverse them—and that's traveling at the speed of light! Eventually, the photon would enter a relatively small group of galaxies (on the outskirts of a 100,000-galaxy supercluster) containing our home, the MWG. Even after entering the MWG, the photon still had tens of thousands of years before it would approach the Sun. Along this last segment of the journey, it encountered even more bizarre conditions. Some of the stars exploded as supernovae that scatter the elements critical to life throughout the galaxy. Others, called neutron stars, are more massive than the Sun but so dense that their sizes are similar to a large US city. These neutron stars spin on their axis at rates exceeding 30 times a second and emit beams of harsh X-rays and gamma rays (just like the massive black holes, but closer to home).

Beyond the bizarre stars, the photon also encountered a myriad of "normal" stars. These range in mass from a few tenths of the Sun up to a few tens of the Sun. The more massive stars burn for a few million years, but the least massive burn for trillions of years!

When this photon finally reached Earth nearly 14 billion years later, more than 95% of all stars that will ever form already existed.[1] In just the MWG, hundreds of billions of Earth-sized planets exist around these stars. The photon changed significantly because space was expanding during the entire journey. The original wavelength near visible light has stretched to microwaves. Furthermore, the location where the photon originated now sits almost 50 billion light-years away! We live in a truly gargantuan universe that has grown and developed for a tremendous amount of time.

When considering the staggering size and age of the universe, many people conclude that life must exist out there. That idea was voiced strongly during the movie *Contact* (1997). The closing scene involves a dialogue between the protagonist Dr. Ellie Arroway and a group of children taking a tour of the Very Large Array (nobody ever said scientists used creative names). In response to a child who asks if there are other people out there in the universe, Ellie says:

> The most important thing is that you all keep searching for your own answers. I'll tell you one thing about the universe, though. The universe is a pretty big place. It's bigger than anything anyone has ever dreamed of before. So, if it's just us, it seems like an awful waste of space.[2]

This assertion resonates with a large number of people. Examining life here on Earth adds weight to the argument. Life abounds on our home planet! Animals fill almost every conceivable niche available—from deserts to mountains, from the surface to the depths of the oceans, from sea level to high mountains, and almost everywhere in between. The smaller the life, the more varied the environments they inhabit. Scientists have discovered bacterial life that flourishes in temperatures both above the boiling point and below the freezing point of water. Other organisms grow optimally in highly basic or highly acidic conditions. Others still thrive in high-radiation environments, extremely dry conditions, or situations with high concentrations of metals, salt, or sugar. Life exists in every possible environment—even in some that scientists once thought impossible!

One final fact to buttress the claim that life exists out there is that life appeared on Earth almost as soon as the planet could support it. As my colleague Fazale Rana points out in his book *Origins of Life*, the fossil record shows life at 3.5 billion years and the geochemical record indicates life as far back as 3.8+ billion years.[3] Considering that asteroids and comets pummeled Earth until about 3.9 billion years ago, life appeared in a geological instant—as soon as Earth could possibly host it.

Might Earth Be the Only Planet to Host Life?
Life requires elements that permit complex biochemical reactions. Carbon seems uniquely suited to fulfill this role. Water serves as the ideal solvent for carbon chemistry. As scientists determined how these elements (carbon, oxygen, and hydrogen) formed in the universe, they discovered that slight changes in the laws of physics would lead to a universe devoid of these building blocks. Yet, they also recognize that the laws of physics produced them in abundance. In this way, our universe seems rather "friendly" to life. Even more so considering that most, if not all, stars host an assortment of planets. But why, if our universe were so friendly to life's building blocks, would one argue that only Earth harbors life?

Carbon, liquid water, and planets are necessary components for life. However, investigations of the planets in our solar system demonstrate that these components are not sufficient to produce life. Studies of Mars show an abundance of evidence for liquid water in the past and transient periods of water more recently. In fact, one set of images revealed two massive debris fields likely caused by tsunamis roughly 3.5 billion years ago.[4] Any liquid water on Mars has long since disappeared,[5] but the ancient ocean required for any tsunami to form might have contained some form of microbial life. Given the minimal atmosphere on Mars, the chemical and radiation environment on the planet would quickly destroy any past or present biological material.[6] The strong evidence of water on Mars in the past provides scientists with a promising opportunity—the chance to study, in far more detail than any exoplanet, the possibility that Mars actually hosted life.

Perhaps probes sent to Mars in the not-too-distant future could excavate the tsunami debris fields and test for any extant or extinct life from the ancient shoreline. Similar opportunities exist with Enceladus and Europa. These moons of Saturn and Jupiter, although they are covered by miles-thick layers of ice, also show strong evidence of liquid water. Maybe future missions could collect the water (as it occasionally escapes from cracks in the ice) and test it for signs of current or past life. If detailed studies of the known liquid water environments in the solar system show no evidence of life, these results would lend credence to the idea that life is a rare phenomenon in the universe.

A growing body of evidence indicates that Earth's remarkable capacity to host a thriving and diverse array of life requires far more than liquid water. Perhaps any habitable planet requires an uncommon migration of planets during the early period of formation, a suitable size for ongoing plate tectonic activity, a just-right sized moon, a suitably dynamic atmosphere that adjusts to the changing brightness of the Sun, and many other characteristics. If so, maybe the odds of a truly habitable planet dwarf the number of planets available. Given these possibilities, Earth may be the *only* planet containing life in the observable universe.

It's an Open Question
Scientifically speaking, the existence of intelligent, sentient life beyond the confines of Earth is an open question. As I survey the scientific landscape, I find that the best explanation for all we see is that the God of the Bible created a universe capable of supporting life, prepared a planet to harbor advanced life, and then created humanity on that planet. The fine-tuned nature of the laws of physics, the purposeful events that could have destroyed Earth but enhanced its ability to support life, and the capacity of humans to understand it all point to something outside of the physical universe that orchestrated the whole story. That said, I think the existence of life beyond Earth is also an open theological question. If God created life here on Earth, why couldn't he have done it somewhere else also?

Let's continue to explore the heavens, searching for distant planets, while developing the technology to determine if life inhabits these remote worlds. I can't help but echo the words of Ellie Arroway when a special committee presses her to provide evidence of her encounter with ET.

> I can't prove it, I can't even explain it, but everything that I know as a human being, everything that I am tells me that

it was real! I was given something wonderful, something that changed me forever. A vision of the universe that tells us, undeniably, how tiny and insignificant and how rare and precious we all are! A vision that tells us that we belong to something that is greater than ourselves, that we are *not*, that none of us are alone! I wish I could share that. I wish that everyone, if even for one moment, could feel that awe and humility and hope.[7]

Ellie Arroway contends that discovering ET will bring hope, vision, and purpose. My wish is that as you examine the heavens and see the wonder and majesty of creation, you come to know that we are not alone in the universe. The Creator, who revealed himself in Jesus Christ, fashioned this universe so that we could have a relationship with him. Indeed, we are not alone and that brings true awe, humility, and hope.

Chapter 7

Why I Believe God Exists: A Biochemical Case for the Creator

by Fazale Rana

Does God exist? What role does science play in answering this vital question? Do scientific advances eliminate the need for a Creator—or do they, in fact, undermine the evolutionary paradigm so often used to justify atheism?

These questions are important for Muslims and Christians alike. If evolutionary mechanisms can account for the origin and history of life and the design of biological systems, then it is right to ask if there is any role for a Creator to play. In his book *The Blind Watchmaker*, evolutionary biologist and atheist Richard Dawkins notes, "Although atheism might have been logically tenable before Darwin, Darwin made it possible to be an intellectually fulfilled atheist."[1]

Statements like this cause many people to conclude that conflict exists between science and religion and science will eventually win the war. In August 2015, the Pew Research Center published data showing that almost 75 percent of Americans who never or seldom attend church believe there is conflict between science and faith. Unfortunately, this same survey showed that 50 percent of regular church attendees believe the same.[2]

And yet, it was science that led me to the strong conviction that a Creator must exist. When I entered college, I was an agnostic. I didn't know if God existed or not, and, honestly, I didn't really care. Religion did not interest me. Instead, my attention centered on biochemistry. I was eager to prepare myself for graduate school so I could earn a PhD in my chosen field.

As an undergraduate, I became convinced that evolutionary mechanisms could account for the origin, history, and design of biological systems. My convictions were not based on a careful examination of the evidence, but rather on what my biology professors taught me. I admired them; thus, I accepted their claims about the evolutionary paradigm without hesitation. In many ways, my misplaced confidence in evolutionary explanations fueled my agnosticism. Today, when I speak on university campuses in the United States, I often encounter students who, like me, embrace the evolutionary paradigm without criticism because they, too, respect and admire their professors.

But my views changed during graduate school. You might say that biochemistry convinced me that God must exist. One of the primary goals of a graduate education is to teach students to think independently through the scientific evidence and to develop conclusions based on the evidence alone, regardless of what other experts might say. Because I was learning to think for myself, I was willing to ask questions that I had not voiced as an undergraduate. The elegance, sophistication, and ingenuity of biochemical systems prompted me to ask, "How did life originate?" I wanted to know how the scientific community could account for the origin of such remarkable systems through strictly mechanistic processes.

After examining the various explanations available at that time (this was 30 years ago), I was shocked. The scientific community's explanations seemed woefully inadequate. I was convinced that chemical and physical processes alone could not generate life. This realization, coupled with the elegant design of biochemical systems, forced me to the conclusion—for intellectual reasons alone—that a Creator must indeed exist and must have been responsible for bringing life into being.

That was over 30 years ago, and in the subsequent decades, the scientific evidence has continued to affirm my convictions about God's existence. The case for a Creator from the design of biochemical systems and the problems associated with the origin of life have become even more compelling.

The goal of this chapter is to present the reasons why, as a biochemist, I think God must indeed exist. My argument can be summarized using three keywords:

1. Fingerprints: A Creator's fingerprints are evident in biochemical systems.
2. Failure: All avenues taken to explain the origin of life through chemical evolution have resulted in failure.
3. Fashion: Attempts to create and fashion life in the lab make a powerful case for a Creator.

Fingerprints

When human beings design, create, and invent systems, objects, and devices, we leave behind telltale signatures in our creations that reflect the work of a mind. As a biochemist, I find it remarkable that the hallmark features of the cell's chemical systems are identical to those features that we would recognize as evidence for the work of a human designer. If specific features reflect the work of human intelligence, and we see those very features in the cell's chemical makeup, then is it not reasonable to conclude that intelligent design undergirds life itself?

Signatures of intelligent design abound in the cell. I simply cannot discuss all of them here. Instead, I would like to focus on the information systems found inside the cell.

Biochemical Information Systems
At their essence, biochemical systems are information systems. Two classes of biomolecules harbor information: (1) the nucleic acids, such as DNA and RNA; and (2) proteins. Both types of molecules are chain-like. These molecules are formed when the cell's machinery links together smaller, subunit molecules in a head-to-tail fashion to form molecular chains. In the case of DNA and RNA, the subunit molecules are called nucleotides. In the case of proteins, the subunit molecules are called amino acids. Twenty different amino acids are encoded within the genetic code. The cell's machinery uses these 20 amino acids to construct proteins.

Biochemists often think of nucleotides and amino acids as molecular alphabets. (Nucleotides are sometimes referred to as the genetic letters abbreviated A, G, C, and T.) Just as alphabet letters are used to construct words in English or Turkish, so amino acid sequences are used to construct biochemical words—proteins—that carry out specific functions inside the cell. Nucleotide sequences are used to store information in DNA. In fact, DNA's chief function is to store information that the cell's machinery uses to build proteins. The regions of the DNA molecule that house this information are called genes.

The recognition that biochemical systems are information systems indicates that a Creator generated life. Why? Because our common experience teaches us that minds generate information. When you receive a text message or when you see a sign along the side of the road, you invariably conclude that someone composed those messages to communicate information to you. In like manner, it is reasonable to conclude that a divine Mind generated the cell's information systems.

Organized Information

But the case for the Creator does not rest on the mere existence of biochemical information. The argument is much more sophisticated. As it turns out, information theorists studying problems in molecular biology conclude that the *structure* of the cell's information systems is identical to the organization of human language. Again, it is not solely the presence of information in the cell, but the way in which that information is organized that points to purposeful design.

One of the most provocative scientific insights I have ever learned relates to the structure and function of biochemical information. Computer scientists and molecular biologists have come to realize that the cell's machinery, which manipulates DNA, literally functions like a computer system at its most basic level of operation. Because this insight is so important to the case for a Creator, I would like to spend more time developing the concept. To do so, we need to consider the work of British mathematician Alan Turing.

Turing Machines and the Cell

One could argue that Turing ranks among the most important scientists of the twentieth century. And yet, until a few years ago, most people did not even know who he was. Recently, Turing's vital work in the effort to crack the Nazi Enigma code during World War II garnered fresh attention when the British government released classified information from that time. It turns out that Turing was a war hero. He, along with other cryptanalysts based at Bletchley Park (50+ miles north of London), designed and operated electromechanical machines used to break Enigma. Historians have estimated that these efforts may have shortened the war by two years and saved about 14 million lives.

As if his contributions to the British war effort were not substantial enough, Turing is also considered by many to be the father of modern computer science. Prior to the war and after it, Turing produced theoretical work that heavily influenced the theoretical framework for modern-day computer science. Much of the technology we enjoy today traces its origin, in part, to theoretical insights that flowed out of Turing's mind.

Part of Turing's theoretical construct for computer systems were abstract machines, known today as Turing machines. These abstract entities existed solely in Turing's mind. Turing machines are very simple, consisting of three components: (1) the input, (2) the finite control, and (3) the output.

The input is a string of data that flows into the finite control. The finite control operates on the data string in a prescribed manner, altering it and generating an output string of data. The finite control can perform only limited operations on the data string. Turing's genius was to recognize that the output of one Turing machine could become the input to another. In this way, an ensemble of Turing machines can be combined in numerous ways to perform numerous, distinct, complex operations.

This is precisely the same way that the cell's machinery manipulates genetic information. Information housed in DNA is digital information. Whenever a complex biochemical process (such as DNA replication) takes place, the cell's machinery—in the form of proteins—takes the digital information in DNA as input, alters it in a prescribed manner, and produces an output strand of digital genetic information. The individual proteins serve as the finite control. While each protein can perform only a limited transformation of the DNA information, by working in combination with other proteins, more complex biochemical operations ensue. In other words, when the cell's machinery replicates DNA, it is essentially carrying out a computer operation.

DNA Computing

The similarity between cellular processes and the fundamental operation of computer systems has inspired a brand-new area of nanotechnology called DNA computing. This cutting-edge field is the brainchild of computer scientist Leonard Adleman of the University of Southern California. Adleman proposes that DNA computing paves the way for a new understanding of life:

> The most important thing about DNA computing is that it shows that DNA molecules can do what we normally think only computers can do. This implies that Computer Science and Biology are closely related. That every living thing can be thought to be computing something, and that, sometimes, we can understand living things better by looking at them as computers.[3]

DNA computers are made up of DNA and the proteins that manipulate this biomolecule. These "biocomputers" are housed in tiny test tubes, yet they are more powerful than the most advanced super computer system we have available to us. That power stems largely from their capacity to perform a vast number of parallel computations simultaneously.

DNA computing highlights the remarkable similarities between human designs and the biochemical designs inside the cell. We can use these astounding similarities to construct a formal argument for God's existence by following in the footsteps of British natural theologian William Paley.

Divine Watchmaker

In 1802, Paley wrote a book called *Natural Theology* in which he advanced the Watchmaker argument, which went on to become one of the best-known arguments in the West for God's existence. In a nutshell, Paley reasoned that just as a watch requires a human watchmaker, so life requires a divine Watchmaker.

In Paley's day, a well-made watch exemplified expert craftsmanship. Paley pointed out that a watch is a contrivance—a machine composed of several parts that interact precisely to accomplish its purpose. He then contrasted a watch with a stone. A stone, Paley argued, finds explanation through the outworking of natural processes; but a watch requires a mind to explain its existence. Based on a survey of biological systems, Paley concluded that living systems have more in common with the watch than with a stone. And if a watch requires a human watchmaker to explain its existence, then by analogy, living systems require a divine Mind to explain their existence.

Advances in biochemistry allow us to bring the Watchmaker argument into the twenty-first century. We know from common experience that computer systems—the pinnacle of engineering achievement in our day—require a mind (in fact, many minds) to explain their existence. And because we find computer systems operating within the cell, we can reasonably conclude that life requires a divine Mind to account for its existence.

I find the Watchmaker argument compelling. Yet in my experience when I present this argument to skeptics they often argue that evolutionary processes can serve as the Watchmaker. In fact, many evolutionists, such as Dawkins, regard these processes as the blind watchmaker. Dawkins articulates this idea in his book:

> [Paley] had a proper reverence for the complexity of the living world, and he saw that it demands a very special kind of explanation. The only thing he got wrong was the explanation itself. ... The true explanation ... had to wait for ... Charles Darwin. ... Natural selection, the blind, unconscious, automatic process which Darwin discovered, and which we now know is the explanation for the existence and apparently purposeful form of all life, has no purpose in mind. It has no mind and no mind's eye. It does not plan for the future. It has no vision,

no foresight, no sight at all. If it can be said to play the role of watchmaker in nature, it is the blind watchmaker.[4]

Dawkins's challenge brings us to the second point of my argument for God's existence.

Failure

To account for the origin of biochemical systems within the evolutionary paradigm, researchers must appeal to a set of processes dubbed chemical evolution, which is the "blind watchmaker" of biochemical systems. But, as I will show, every attempt to explain the genesis of biochemistry via chemical evolution has resulted in frustration and failure.

To appreciate how severe this problem is, we need to review the way in which biochemists categorize the different types of biochemical systems. Biochemists frequently organize biochemical systems into three categories: (1) information-rich molecules (proteins and nucleic acids), (2) intermediary metabolism, and (3) cell membranes. (We have already discussed information-rich biomolecules such as proteins and nucleic acids.)

Intermediary metabolism refers to the collection of chemical reactions involving small molecules that take place inside the cell. These chemical reactions are organized into a series of pathways in which one molecule is converted into another molecule. The pathways can be linear, branched, or cyclical. Often, the pathways will intersect with each other to form a highly reticulated ensemble of chemical reactions. These chemical pathways are critical for harvesting energy for the cell's use. They are also used to make the building blocks that form DNA, RNA, proteins, and the constituents of cell membranes, and to process molecular waste secreted from the cell.

Cell membranes form boundaries that separate the interior of the cell from the exterior environment. Membranes also form compartments within the cell's interior.

Chemical Evolution Models

These three categories of biochemical systems have inspired corresponding models for chemical evolution: (1) replicator-first, (2) metabolism-first, and (3) membrane-first.

According to replicator-first scenarios for the origin of life, the first biochemical systems centered on information-rich molecules and it was only later in the evolutionary process that metabolism and cell membranes emerged. In like manner, metabolism-first scenarios envision metabolic systems emerging first and, finally, for membrane-first scenarios cell membranes arise first.

It is important to realize that each approach to the origin of life suffers from intractable problems. For example, the replicator-first approach suffers from the monomer problem, the homochirality problem, and the homopolymer problem.[5]

Many origin-of-life researchers readily acknowledge these profound issues. In 2002, I attended the International Society for the Study of the Origin of Life in Oaxaca, Mexico. The conference attracted some of the best origin-of-life scientists, including the late Leslie Orgel. When he was alive, Orgel was considered the preeminent origin-of-life researcher in the world. Because of his status, Orgel was given the honor of opening the conference with a lecture in which he offered his perspective on the RNA world hypothesis. This idea, which Orgel himself conceived, is considered the most important idea in origin-of-life research. Yet throughout his lecture, Orgel detailed problem after problem with the RNA world hypothesis. Toward the end of his talk, he paused, then said, "I hope that there are no creationists in the audience, but it would be a miracle if a strand of RNA ever appeared on the primitive Earth."[6]

It was remarkable. Orgel was known as an outspoken atheist, yet in an honest moment he acknowledged that the origin of life requires a miracle—at least when envisioned as a replicator-first scenario. As I point out in my book *Creating Life in the Lab*, Orgel's conclusion still holds today.

Metabolism-First Models
Metabolism-first scenarios fare no better. They are susceptible to disruption due to chemical interference and proceed too slowly to be effective means of generating life.

The scenario is also dependent on transport mechanisms and on mineral catalysts (but mineral surfaces have limited catalytic range). Before Orgel died, he described metabolism-first scenarios as "an appeal to magic," "a series of remarkable coincidences," and "a near miracle."[7] Again, Orgel's conclusion still holds. (See *Creating Life in the Lab*.)

Membrane-First Models
And finally, membrane-first scenarios are also riddled with problems. These models require exacting environmental conditions, amphiphile composition, amphiphile concentration, and phase behavior. Plus, the requirements for each step in these scenarios are incompatible.

A few years ago, chemist Jackie Thomas and I published a paper detailing the problems with membrane-first scenarios, in the journal *Origins of Life and Evolution of Biospheres*. In my view, this was a remarkable achievement. Both Thomas and I are creationists. And yet the problems we identified with membrane-first scenarios are so significant, even evolutionary biologists had to acknowledge that our critique was legitimate. The editor-in-chief of the journal consented to publish our critical assessment of membrane-first scenarios in the premier origin-of-life publication.

In other words, every attempt to explain the origin of life has resulted in failure. There is no blind watchmaker.

Honest skeptics will agree that—as of now—we do not have an explanation for the origin of life. However, in my experience, when presented with critiques of chemical evolution scenarios, skeptics often argue that chemists have identified numerous chemical and physical processes in the laboratory that could conceivably contribute to the origin of life. They claim that these insights constitute important clues as to how life emerged through chemical evolution. For example, they will highlight such laboratory successes as:

- the synthesis of most "building block" molecules
- production of biopolymers
- evolution of functional RNA molecules
- generation of self-replicating systems
- manufacture of protocells

This response leads us to the third component of our argument for God's existence.

Fashion

When chemists go into the lab and perform prebiotic chemistry studies, they are working under highly controlled conditions. They carefully assemble the glassware and fill it with the appropriate solvents. They add the just-right chemicals at the just-right time and just-right concentrations. They control the reaction's temperature and pH. And they stop the reaction at that just-right time.

In other words, the chemists are contributing to the success of the prebiotic chemistry studies. It is highly questionable whether such tightly controlled conditions existed on early Earth. To put it another way, intelligent agency ensures the success of prebiotic reactions in the lab.

Let me illustrate this point by discussing the RNA world hypothesis. The centerpiece of this idea is the notion that the very first biochemistry on Earth was RNA-based. (Later, the RNA world evolved to give rise to the DNA/protein world that characterizes contemporary biochemistry.) Origin-of-life researchers highlight several lines of evidence in favor of the RNA world hypothesis. I'm going to discuss just one of them.

Creating RNA Molecules in the Lab

In the mid-1990s, scientists in the lab observed RNA molecules assembling on clay surfaces from chemical building blocks. This observation was heralded as a huge breakthrough for the RNA world hypothesis. It meant that RNA could conceivably form on early Earth with an assist from clays and other minerals. Yet when the RNA assembly experiments are examined, it becomes evident that intelligent agency was critical to ensure the success of this chemical process.

For example, the researchers who performed the studies excluded materials that would interfere with RNA assembly on clay. These disruptive materials most certainly would have been present on early Earth. Researchers were also very careful to exclude materials that would promote the breakdown of RNA molecules. Such materials most certainly would have been present on early Earth. The team also stopped the reaction before the RNA molecules got too long and became irreversibly attached to the clay surface. When that happens, the RNA molecules will no longer be available for subsequent steps in the origin-of-life process. There would not have been organic chemists on early Earth to stop the chemical reaction before the RNA became permanently bound to the clay.

Additionally, the researchers had to use chemically activated building blocks to make the RNA molecules on clay. These activated building blocks would not have been present on early Earth. And if they were, they would be so reactive that they would have reacted with all kinds of materials before they could form a strand of RNA. Lastly, the scientists had to buy the clay they used in their experiments from a specific supplier and then still had to treat the clay in the lab to remove all the ions except for sodium.

It is ironic that the very experiments that have been performed to demonstrate that chemical evolution could explain the origin of life drive us to the conclusion that intelligent agency is required for life to originate from a complex chemical mixture. Astrobiologist Paul Davies had this to say in response to these experiments:

As far as biochemists can see, it is a long and difficult road to produce efficient RNA replicators from scratch. . . . This conclusion has to be that, without a trained organic chemist on hand to supervise, nature would be struggling to make RNA from a dilute soup under any plausible prebiotic conditions.[8]

Evolutionary biologist Simon Conway Morris said:

> . . . many of the experiments designed to explain one or other step in the origin of life are either of tenuous relevance to any believable prebiotic setting or involve an experimental rig in which the hand of the researcher becomes for all intents and purposes the hand of God.[9]

These conclusions are affirmed by work in the relatively new discipline called synthetic biology.

Creating Life in the Lab
One of the goals of synthetic biology is to construct artificial cells in a laboratory setting. When one examines the work in synthetic biology, it becomes apparent the necessary role intelligent agency plays in transforming simple chemical materials into protocells. Let me illustrate this point by looking at what it takes to make a single enzyme, which would serve as one minor component in the cell's machinery.

A few years ago, a research team sought to make an enzyme that performed a chemical reaction not found in biological systems. In addition to biochemists and molecular biologists, the project needed a team of quantum chemists, computational chemists, and protein engineers. It required not just these skilled scientists, but also hundreds of hours of supercomputer time, use of structural motifs from proteins in nature, and sophisticated instrumentation.

The work of these researchers can rightly be considered science at its very best. And yet when their enzyme was compared to the enzymes typically found in biological systems, their accomplishment was laughable.

> Although our results demonstrate that novel enzyme activities can be designed from scratch and indicate the catalytic strategies that are most accessible to nascent enzymes, there is still a significant gap between the activities of our designed catalysts and those of naturally occurring enzymes.[10]

Divine Watchmaker Revealed
It seems the Watchmaker isn't blind at all. To summarize, we made a case for God's existence by showing that:

1. A Creator's fingerprints are evident in biochemical systems;
2. All avenues taken to explain the origin of life through chemical evolution have resulted in failure; and
3. Attempts to create and fashion life in the lab make a powerful case for a Creator.

It is gratifying to me that the reasons that convinced me to believe in a Creator 30 years ago are still valid today. In my view, if one is truly open to the evidence at hand, there is only one conclusion: a Creator must exist and must be responsible for bringing the very first life-forms into existence. And if a Creator is responsible for the origin of life, then it is reasonable to think that the history of life stems from that Creator's handiwork as well.

Chapter 8

The Case for the Image of God

by Fazale Rana with Kenneth Richard Samples

What would happen if Iron Man (Tony Stark) squared off with Dr. Octopus, one of Spider-Man's most dangerous and notorious foes? In an experiment gone awry, brilliant physicist Otto Octavius accidentally permanently fused a set of highly advanced mechanical arms—controlled via a brain-computer interface—to his body and brain. This accident transforms the weak and nearsighted Octavius into the dangerous and imposing figure Dr. Octopus, complete with a harness of indestructible arms.

Iron Man takes on Dr. Octopus in the *Unfixable* story arc (*The Invincible Iron Man* #501–503), written by Matt Fraction. But the combat isn't physical. It's a battle between two brilliant minds. Obsessed with touting his own genius, Octavius was insulted and dismissed by a young, drunken Stark years earlier when the two of them attended the same technology conference. And Octavius never forgave Stark.

Dying from brain damage, Octavius wants one final chance to humiliate Stark. He poses a mental challenge that he forces Stark to accept by threatening to detonate a 21-kiloton nuclear device over Manhattan and by holding hostage Stark's young protégé, Timothy Cababa. Dr. Octopus wants Stark to "fix" him or admit that he is "unfixable." He tells Stark, "Simply admit it is entirely too complicated a condition for you to solve and it all goes away. I want to see you say, 'I can't fix you.'"

With the timer on the nuclear device counting down toward zero, Stark has only one option: to acknowledge he can't fix Dr. Octopus. To keep the bomb from detonating, Stark gets down on his knees, admitting that Dr. Octopus is smarter than him.

The *Unfixable* story arc highlights the real source of Stark's power as a

ıot the Iron Man armor he dons, but his mind. Tony Stark's
him apart. It makes him one of a kind.

Humans Are Exceptional

The same is true for human beings. The source of our strength is our minds.
It makes us the planet's most powerful and imposing creatures. This feature
provides us with the means to produce human enhancement technology and
possibly one day take control of our own "evolution." Our advanced cognitive
capabilities appear to set us apart from all life on Earth. For Christians, human
advanced cognitive ability reflects an aspect of the image of God—a special
quality possessed by each and every human being, endowed to humanity at the
point of our creation.

Human Evolution and Human Nature

In the same way that Octavius sought to denigrate Stark's intellectual capabilities, many in the scientific community wish to do the same to the biblical
concept of humanity. These scientists view human beings as products of the
unremarkable outworking of evolutionary processes, rather than the crown
of God's creation. And though we might think of ourselves as exceptional, we
aren't—at least according to the evolutionary paradigm. We are just one among
countless accidental species that have existed throughout Earth's history.

Concerned that they will be accused of speciesism, these scientists regard
"exceptional" qualities as having antecedents in the behavioral capacities of
other creatures such as the great apes and hominins (for example, *Homo erectus*
and Neanderthals) that preceded us in the fossil record. In other words, human
beings differ in *degree*, not *kind*, from other creatures. It is a battle between two
worldviews over the origin and nature of humanity—and of our minds.

An evolutionary perspective of humanity's origin and nature has dominat-
ed anthropology since the publication of Darwin's famous work *The Descent
of Man, and Selection in Relation to Sex*. While Darwin left the topic of human
origins unaddressed in *On the Origin of Species*, he wrote about it in detail in
The Descent of Man. Darwin argued that, like all species, humanity evolved
through a process of descent with modification from an ancestor shared with
apes. As Darwin put it, "In a series of forms graduating insensibly from some
ape-like creature to man as he now exists, it would be impossible to fix on any
definite point when the term 'man' ought to be used."[1]

Darwin posited that human beings are nothing more than animals, not
the product of divine activity. In his day, Darwin had done the unthinkable;

he interpreted human history in a fully mechanistic and materialistic fashion. According to this view, all human nature, not just humanity's physical makeup, emerged under natural selection's auspices. Darwin regarded not only humanity's mental powers and intellectual capacity but also our moral sense and religious beliefs as evolution's creation.

The central ideas of human evolution extend beyond the "ivory towers" of the academy to influence much of our culture. Human evolution has become the modern-day creation myth in many cultures around the world. As a result, many people reject the biblical account and embrace a secular view of humanity. For this reason, most ethical deliberations in today's world are secular in orientation and rely on some form of utilitarianism and consequentialism. Yet, a secular approach to ethics does little to offer genuine guidance regarding how advances in biotechnology and bioengineering should be used.[2]

In the face of the inadequacy of secular bioethical systems, what are people to do? Is there an ethical framework that *can* provide the guidance we desperately desire? As Christians, we believe that Christianity can provide such an ethical framework. We believe a system of bioethics based on the Christian worldview has the capacity to successfully guide deliberations about which biotechnologies should be developed and how they should be deployed.[3] This framework finds its basis in the notion that human beings are created in God's image. Yet, the prominence of human evolution causes many people to believe that this idea is outmoded.

However, as a philosopher and a scientist, we believe a strong case can be made for the image of God, *apart from Scripture*. Because of the centrality of the image of God to a biblically based ethical system, we maintain that it is critical to justify this view of humanity before we use it to anchor an ethical system. Toward this end, we devote this chapter to the case for human exceptionalism, and hence, the image of God.

What Is the Image of God?

Before we present a defense of this key biblical concept, it is important that we define what we mean by the image of God. This is no simple undertaking. Scripture doesn't explicitly state what the image of God is, but it does provide some important clues. Over the centuries, theologians have discussed and debated this concept. Some take the image of God to describe humanity's spiritual—but finite and limited—resemblance to God. The *resemblance view* has been the historic Christian view. But today the view has fallen out of favor and has been supplanted by two competing interpretations: the *relational view* and the *representative view.*

Some scholars argue that the image of God doesn't define qualities we possess as human beings, but instead refers to the unique relationship human beings can enter into with God. Others argue that the image of God involves responsibilities God has for humanity. Hence, we are to serve as God's representatives or viceroys on Earth.[4] Clearly, these two views find support in the Genesis creation accounts.

Even though theologians can't settle on what the image of God means, two New Testament passages (Colossians 3:10 and Ephesians 4:24) shed important light on the debate. In these passages, Paul encourages the Christians at Colossae and Ephesus to allow the Holy Spirit to transform them into the image of their Creator. These passages imply that God's image includes our capacity for knowledge, understanding, love, holiness, and righteousness. In other words, according to Paul, the image of God refers to attributes we possess as humans. We think these passages affirm the historic view and, therefore, tip the scales in favor of the resemblance view.

Still, it is worth noting that the three perspectives on the image of God aren't mutually exclusive. In fact, we would argue that to serve as God's representatives on Earth and to enter into a relationship with our Maker requires that we possess attributes that resemble God's, at least in some measure. Another way to put it: because we resemble God—at least in part—we can be granted the responsibility to function as his vice regents on Earth and we have the privilege to uniquely enter into a special type of relationship with God. For readers who want a more detailed discussion on the image of God, see either *A World of Difference* or *7 Truths That Changed the World,*[5] both written by Kenneth Samples.

Working toward a consensus of the three approaches, we identify four characteristics that give parameters to the concept of God's image.[6]

1. Human beings possess a moral component. We inherently understand right and wrong and have a strong, innate sense of justice.
2. Humans are spiritual beings who recognize a reality beyond this universe and physical life. We intuitively acknowledge the existence of God and have a propensity for worship and prayer. We desire to connect to the transcendent.
3. Human beings relate to God, to ourselves, to other people, and to other creatures. There is a relational aspect to God's image.
4. Humanity's mental capacity reflects God's image. Human beings

possess the ability to reason and think logically. We engage in symbolic thought. We express ourselves with complex, abstract language. We are aware of the past, present, and future, and we display advanced creativity through art, music, literature, science, and technical inventions.

As Christians, we believe these qualities set us apart from all other creatures; they make us exceptional. And we think these qualities can serve as the foundation for a robust ethical system that can guide the pursuit and usages of human enhancement technologies. We believe that our convictions about humanity's nature find independent corroboration from philosophy and science, as the following pages attest.

The Case for Human Exceptionalism

The notion that human beings differ in degree, not kind, from other creatures has been a mainstay concept in anthropology and primatology for over 150 years. And it has been the primary reason why so many people have abandoned the belief that human beings bear God's image. Yet this stalwart view in anthropology is losing its mooring, and the concept of human exceptionalism is taking its place. Remarkably, a growing minority of anthropologists and primatologists—steeped within the evolutionary paradigm—now believe that human beings really are exceptional. They contend that human beings do, indeed, differ in kind, not just degree, from other creatures. Scientists who argue for this updated perspective have developed evidence for human exceptionalism within the context of the evolutionary paradigm in their attempts to understand how the human mind evolved. Ironically, the new insights marshal support for the biblical conception of humanity and raise challenging questions for the evolutionary account.

These anthropologists and primatologists identify at least four interrelated qualities that make us exceptional: (1) symbolism, (2) open-ended generative capacity, (3) theory of mind, and (4) capacity to form complex social networks. From our perspective as Christians, we consider these qualities as scientific descriptors of the image of God.

Human beings effortlessly represent the world with discrete symbols, and we denote abstract concepts with symbols. Our ability to represent the world symbolically has interesting consequences when coupled with our ability to combine and recombine those symbols in uncountable ways to create alternate possibilities. According to Thomas Suddendorf, a psychologist who studies mental development in humans and animals:

A key to our open-ended, generative capacity is our ability to recursively embed one thing in another, as it enables us to combine and recombine basic elements such as people, objects, and actions into novel scenarios. Such nesting is also essential for reflection: our capacity to think about our own thinking. Nested thinking allows us to reason about the mental scenarios we entertain. . . . We can connect diverse scenarios into larger plots. . . . We can reflect on the relationship between past experiences and construct complex plans with embedded if-then steps.[7]

Human capacity for symbolism manifests in the form of language, art, music, and body ornamentation. And we desire to communicate the scenarios we construct in our minds with other human beings. But we long for more than to merely communicate. We want to link our minds together. We can do this because we possess theory of mind. In other words, we recognize that other people have minds just like ours, which allows us to understand what others are thinking and feeling. We also have the brain capacity to organize people we know into hierarchical categories that help us form and engage in complex social networks.

An interesting illustration of our capacity to link our minds together comes from a recent study by Spanish neuroscientists from the Basque Center on Cognition, Brain, and Language. According to this study, when human beings engage in conversations with one another—even with strangers—the electrical activities of our brains synchronize.[8]

Prior to the neuroscientists' investigation, most brain activity studies focused on individual subjects and their responses to single stimuli. For example, single-person studies have shown that oscillations in electrical activity in the brain couple with speech rhythms when the test subject is either listening or speaking. The Spanish neuroscientists wanted to go one step further. They wanted to learn what happens to brain activities when two people engage one another in conversation.

To find out, they assembled 15 pairs (14 men and 16 women) consisting of strangers who were 20 to 30 years in age. They asked the members of each dyad to exchange opinions on sports, movies, music, and travel. While the strangers conversed, the researchers monitored electrical activities in the brains using EEG technology. As expected, they detected coupling of brain electrical activities with the speech rhythms in both speaker and listeners. But to their surprise, they also detected pure brain entrainment in the electrical activities of the test subject, independent of the physical properties of the sound waves associated with speaking and listening. To put it another way, the brain activities of the two people in the conversation became synchronized, establishing a connection between their minds.

It is noteworthy that all four qualities that make human beings exceptional were on full display in the study. The capacity to offer opinions on a wide range of topics and to communicate ideas with language reflects our symbolism and our open-ended generative capacity. It is also intriguing that the oscillations of our brain's electrical activity couple with the rhythmic patterns created by speech, suggesting our brain is hardwired to support our desire to communicate with one another symbolically. It is equally intriguing that our brains become coupled at an even deeper level when we converse, consistent with our theory of mind and human capacity to enter into complex social relationships.

According to Suddendorf, these capacities make human beings different in kind from the great apes, creating a chasm that distinguishes humans from other creatures. But aren't these capabilities present in lesser forms in other animals? Suddendorf argues that the behaviors displayed by animals that many anthropologists have traditionally interpreted as antecedents to human capabilities are fundamentally different from the behaviors we possess. To put it more concretely:

- animal communication differs in kind from open-ended human language,
- animal memory differs in kind from human mental time travel,
- problem-solving in animals differs in kind from human abstract reasoning,
- empathy displayed by animals differs in kind from morality, and
- social cognition in animals differs in kind from the cumulative culture possessed by human beings.[9]

What about Neanderthals?
Although a growing number of anthropologists and primatologists regard humans as different in kind from the great apes, they maintain that the capabilities that separate us from other creatures, such as symbolism, open-ended generative capacity, and theory of mind, emerged through an evolutionary process. Thus, they argue that hominins such as Neanderthals or *Homo erectus* possessed a proto-symbolism that gave rise to sophisticated symbolism found

in human cultures.

In the case of *H. erectus*, the archaeological record fails to yield any evidence for proto-symbolism. As discussed in *Who Was Adam?* (a book Fazale Rana coauthored with Hugh Ross), the singular claim that *H. erectus* displayed symbolic capacities fails to withstand scientific scrutiny.[10]

But what about Neanderthals? Based on archaeological and fossil finds, some paleoanthropologists argue that these hominids: (1) buried their dead, (2) made specialized tools, (3) used ochre, (4) produced jewelry, (5) created art, and (6) had language capacities. If these claims are true, they undermine the notion of human exceptionalism. Such behaviors overlap with ours, meaning that human beings can't be viewed as unique.

These claims also lend credence to the notion that our exceptional capabilities weren't created, but rather emerged through an evolutionary history. Paleoanthropologists believe that humans share an evolutionary ancestor with Neanderthals. If so, these scientists would argue that similar cognitive capacities imply that the ancestral creature that gave rise to human and Neanderthal lineages must have possessed these capabilities as well—or at least precursors to them.

But, as we argue in *Who Was Adam?* (and elsewhere), none of these claims withstand ongoing scientific scrutiny.[11] If that is so, why do claims about Neanderthal symbolism abound? Since their discovery, Neanderthals have been considered by most paleoanthropologists to be cognitively inferior to human beings. Claims of Neanderthal symbolism are a relatively recent phenomenon. Science writer Jon Mooallem argues that paleoanthropologists have been slow to acknowledge the sophisticated behavior of Neanderthals due to bias that reflects the earliest views about these creatures—namely, that they are "unintelligent brutes."[12] Accordingly, this view has colored the way paleoanthropologists interpret archaeological finds associated with Neanderthals and keeps them from seeing the obvious: Neanderthals had sophisticated cognitive abilities. In fact, Mooallem accuses paleoanthropologists who continue to reject this new view of Neanderthals as being "modern human supremacists," guilty of speciesism borne out of an "anti-Neanderthal prejudice."[13]

Mooallem offers a reason why this prejudice continues to persist among some paleoanthropologists. In part, it is because of the limited data available to them from the archaeological record. In the absence of a robust data set, paleoanthropologists must rely on speculation fueled by preconceptions. Mooallem states:

All sciences operate by trying to fit new data into existing theories. And this particular science, for which the "data" has always consisted of scant and somewhat inscrutable bits of rock and fossil, often has to lean on those metanarratives even more heavily. . . . Ultimately, a bottomless relativism can creep in: tenuous interpretations held up by webs of other interpretations, each strung from still more interpretations. Almost every archaeologist I interviewed complained that the field has become "overinterpreted"—that the ratio of physical evidence to speculation about that evidence is out of whack. Good stories can generate their own momentum.[14]

Yet, as we have pointed out, careful examination of the archaeological and fossil evidence reveals the speculative nature of Neanderthal "exceptionalism" claims. Could it be that claims of Neanderthal language, art, and religion result from an overinterpreted archaeological record, and not the other way around?

In effect, Mooallem's critique of the "modern human supremacists" cuts both ways. In light of the limited and incomplete data from the archaeological record, it could be that paleoanthropologists who claim Neanderthals show sophisticated cognitive capacities, just like modern humans, have their own prejudices. Their view is fueled by an "anti-modern human bias"—one that seeks to undermine modern humans' uniqueness and exceptionalism—and a speciesism all their own. And to do this they must make the claim that Neanderthals were just like us.

Perhaps if we examined other data beyond the archaeological record, we would gain a better understanding of Neanderthals' cognitive abilities. What do studies of Neanderthal brain structure, development, and genetics tell us about their behavioral capacities?

As detailed in *Who Was Adam?*, the Neanderthal brain *shape* differed from ours, though their brain *size* was about the same as that of modern humans. For example, Neanderthals appear to have had an underdeveloped parietal lobe compared to modern humans.[15] In humans, this brain area plays a role in language, math reasoning, sense of self, and religious experiences. Paleoanthropologists have also come to learn that Neanderthals had a smaller area of their brain devoted to keeping track of social networks than do modern humans.[16] These two results indicate that Neanderthal cognitive capacity was inferior to ours.

Differences between developmental trajectories of Neanderthals and

---

The Origin of Language and Human Exceptionalism

In our view, the one capability that exemplifies our exceptional nature as human beings is our capacity for language. Researchers now recognize that language appears suddenly and coincides uniquely with modern humans. Also, the very first language employed by modern humans was as complex as contemporary languages.

Two statements from recent scientific articles illustrate this insight. A researcher at MIT in the United States writes, "The hierarchical complexity found in present-day language is likely to have been present in human language since its emergence."[19] Elsewhere, another research team concedes: "By this reckoning, the language faculty is an extremely recent acquisition in our lineage, and it was acquired not in the context of slow, gradual modification of preexisting systems under natural selection but in a single, rapid, emergent event that built upon those prior systems but was not predicted by them. . . . The relatively sudden origin of language poses difficulties that may be called 'Darwin's problem.'"[20]

The sudden appearance of complex language is exactly what we would expect if human beings are the product of the Creator's hand, made in God's image.

---

modern humans affirm this conclusion. As described in *Who Was Adam?*, based on skull structures and tooth microanatomy, paleoanthropologists believe that Neanderthals developed from childhood to adulthood more rapidly than we do.[17] Our protracted time in adolescence plays a key role in human brain development. Without this opportunity, it is unlikely that Neanderthals would have had advanced cognition.

Comparisons of the Neanderthal and human genomes also provide insight into our cognitive differences. Researchers have noted differences in genes that play a role in brain development, suggesting that cognitive differences may well exist between modern humans and Neanderthals.[18]

While not definitive, the biological, developmental, and genetic differences measured between humans and Neanderthals support the view that these creatures were cognitively inferior to us, most likely lacking symbolic and open-ended generative capacities. This conclusion doesn't mean that these creatures weren't remarkable in their own right.

Human Enhancement Technology, Transhumanism, and the Image of God
Ironically, progress in human enhancement technology and the prospects of a posthuman future serve as one of the most powerful arguments *for* human exceptionalism and, consequently, the image of God. Human beings are the only species that exists—or that has ever existed—that can create technologies to enhance our capabilities beyond our biological limits. We alone work toward effecting our own immortality, take control of evolution, and look to usher in a posthuman world. These possibilities stem from our unique and exceptional capacity to investigate and develop an understanding of nature (including human biology) through science and then turn that insight into technology. Our exceptional nature is no more evident than when we compare our technological achievements to those produced by the great apes such as chimpanzees and bonobos.

Thomas Suddendorf puts it this way:

> We build machines that speed us from one place to the other, even to outer space. We investigate nature and rapidly accumulate and share knowledge. We create complex artificial worlds in which we wield unheralded power—power to shape the future and power to destroy and annihilate. We reflect on and argue about our present situation, our history, and our destiny. We envision wonderful, harmonious worlds as easily as we do dreadful tyrannies. Our powers are used for good as they are for bad, and we incessantly debate which is which. Our minds have spawned civilizations and technologies that have changed the face of the Earth, while our closest living animal relatives sit unobtrusively in their remaining forests. There appears to be a tremendous gap between human and animal minds.[21]

From an evolutionary perspective, humans and chimpanzees share an evolutionary ancestor about 6 to 7 million years ago. Accordingly, one lineage spawned human beings and the other chimpanzees and bonobos. Because we uniquely possess the capacity for symbolism, open-ended generative imaginations, theory of mind, and the capacity to form complex social networks, humans have launched a succession of ever-increasingly sophisticated technologies. Human enhancements typify our exponentially explosive list of accomplishments. In contradistinction, our counterparts among the great apes display crude and stagnant technology that is of little use to them as they are helplessly being driven to extinction.

Human and Neanderthal technology also appear to have taken different trajectories. Again, this difference seems to reflect our unique and exceptional qualities as human beings. Recently, paleoanthropologist Ian Tattersall and linguist Noam Chomsky (along with other collaborators) argued that Neanderthals could not have possessed language and, hence, symbolism, because their crude "technology" remained stagnant for the duration of their time on Earth. Neanderthals—who first appear in the fossil record around 250,000 to 200,000 years ago and disappear around 40,000 years ago—existed on Earth longer than modern humans have. Yet, our technology has progressed exponentially, while Neanderthal technology remained largely static (as did the technology of other hominids, such as *Homo erectus*). According to Tattersall, Chomsky, and their coauthors:

> Our species was born in a technologically archaic context, and significantly, the tempo of change only began picking up after the point at which symbolic objects appeared. Evidently, a new potential for symbolic thought was born with our anatomically distinctive species, but it was only expressed after a necessary cultural stimulus had exerted itself. This stimulus was most plausibly the appearance of language. . . . Then, within a remarkably short space of time, art was invented, cities were born, and people had reached the moon.[22]

In effect, these researchers echo Suddendorf's point. The gap between humans and the great apes and hominins becomes most apparent when we consider the remarkable technological advances we've made during our tenure as a species. This mind-boggling growth in technology points to our exceptionalism as a species and affirms the biblical view that humans alone bear God's image.

In short, it is possible to make a case—apart from theological considerations—for human exceptionalism from comparative primatology and the archaeological record. Humans really do appear to be different in kind—in a way that comports with the image of God. In a secular world, the scientific evidence for human exceptionalism justifies the use of this theological concept to build an ethical system. As Wesley J. Smith (author and a senior fellow at the Discovery Institute's Center on Human Exceptionalism) points out, "A belief in human exceptionalism . . . does not depend on religious faith. Whether we were created by God, came into being through blind evolution, or were intelligently designed, the importance of human existence can and should be supported by the rational examination of the differences between us and all other known life forms."[23]

Notes

Introduction
by George Haraksin

1. For more on factual evidence and Jesus's epistemology, see Douglas Groothuis, *On Jesus*: *Wadsworth Philosopher Series* (Boston, MA: Cengage Learning, 2002), 51–63.
2. See also Mark 13:26, Matthew 16:21, Luke 24:13–49, John 10:38, and 1 John 1:1.
3. Groothuis, *On Jesus*, 51.
4. Cary Funk and David Masci, "5 Facts about the Interplay between Religion and Science," Pew Research Center (October 22, 2015), pewresearch.org/fact-tank/2015/10/22/5-facts-about-the-interplay-between-religion-and-science/.
5. Dallas Willard, *The Allure of Gentleness: Defending the Faith in the Manner of Jesus* (New York: HarperCollins, 2015), 2.

Chapter 1 – Isn't Faith Incompatible with Reason?
by Kenneth Richard Samples

1. Christopher Hitchens, on "Holier Than Thou," *Penn & Teller: Bullsh*t!*, season 3, episode 4, aired May 23, 2005, imdb.com/title/tt0771115/.
2. Richard Dawkins, "Lions 10, Christians Nil," *New Humanist*, June 1992, reader.exacteditions.com/issues/64700/page/4.
3. See Michael Kruger, "Are Christians Ignorant, Uneducated, Simpletons? Sort Of," *Canon Fodder*, June 25, 2018, michaeljkruger.com/are-christians-ignorant-uneducated-simpletons-sort-of/.
4. Dawkins, "Lions 10, Christians Nil."
5. For more about how historic Christian thinkers have viewed the relationship between faith and reason, see Ed L. Miller, "Faith and Reason," chap. 7 in *God and Reason: An Invitation to Philosophical Theology*, 2nd ed. (Upper Saddle River, NJ: Prentice Hall, 1995).
6. Geoffrey W. Bromiley, gen. ed., *The International Standard Bible Encyclopedia*, vol. 2 (Grand Rapids, MI: Eerdmans, 1982), s.v. "faith," 270.
7. See Kenneth Richard Samples, *A World of Difference: Putting Christian Truth-Claims to the Worldview Test* (Grand Rapids, MI: Baker Books, 2007), 80–81.
8. For many within the historic Christian tradition, saving faith is viewed distinctly as the gift of God (Acts 13:48; 1 Corinthians 12:3; Ephesians 2:8–9; Hebrews 12:2). For a helpful

biblical discussion of how faith is a sovereign gift of God, see Anthony A. Hoekema, *Saved by Grace* (Grand Rapids, MI: Eerdmans, 1989), 143.

9. See Kenneth Richard Samples, *Christianity Cross-Examined: Is It Rational, Relevant, and Good?* (Covina, CA: RTB Press, 2021), chap. 5.

10. See Samples, *Christianity Cross-Examined*, chap. 5.

11. These approaches to faith and reason are explored and explained in Miller, *God and Reason*, 129–153, and in Kenneth D. Boa and Robert M. Bowman Jr., *Faith Has Its Reasons: Integrative Approaches to Defending the Christian Faith*, 2nd ed. (Waynesboro, GA: Paternoster, 2006).

12. Miller, *God and Reason*, chap. 7.

13. See also J. P. Moreland, *Christianity and the Nature of Science: A Philosophical Investigation* (Grand Rapids, MI: Baker Books, 1989).

14. Patrick J. Hurley, *A Concise Introduction to Logic*, 8th ed. (Belmont, CA: Wadsworth, 2003).

15. For Christianity's influence on science, see R. Hooykaas, *Religion and the Rise of Modern Science* (London: Chatto & Windus, 1972); Nancy R. Pearcey and Charles B. Thaxton, *The Soul of Science: Christian Faith and Natural Philosophy* (Wheaton, IL: Crossway, 1994); James Hannam, *The Genesis of Science: How the Christian Middle Ages Launched the Scientific Revolution* (Washington, DC: Regnery, 2011).

16. See Samples, *Christianity Cross-Examined*, chap. 2.

17. See Samples, *Christianity Cross-Examined*, chap. 10.

18. For more on some of Christianity's A-Team, see Kenneth Richard Samples, *Classic Christian Thinkers: An Introduction* (Covina, CA: RTB Press, 2019), 80–81.

19. Gary Habermas, *Dealing with Doubt* (Chicago: Moody Press, 1990), made available for free by the author here: garyhabermas.com/books/dealing_with_doubt/dealing_with_doubt.htm.

20. Abductive reasoning is further explored in Samples, *Christianity Cross-Examined*, chap. 12.

21. Richard Swinburne, *Is There a God?* (New York: Oxford University Press, 1997), 2.

22. Dawkins, "Lions 10, Christians Nil."

Chapter 2 – If God Created Everything, Then Who Created God?
by Kenneth Richard Samples

1. Stephen Hawking, *A Brief History of Time* (New York: Bantam Books, 1998).

2. Seth Stephens-Davidowitz, "Googling for God," *New York Times*, September 19, 2015, nytimes.com/2015/09/20/opinion/sunday/seth-stephens-davidowitz-googling-for-god.html.

3. Bertrand Russell, *Why I Am Not a Christian: And Other Essays on Religion and Related Subjects* (New York: Touchstone, 1967), 6.

4. Richard Dawkins, *The God Delusion* (Boston: Houghton Mifflin, 2006), 188.

5. Daniel Dennett, *Darwin's Dangerous Idea: Evolution and the Meaning of Life* (New York: Touchstone, 1995), 71.

6. Thomas Aquinas, "Question 2. The Existence of God," in *Summa Theologica* (Notre Dame, IN: Christian Classics, 1981), 13.

7. David Hume, in Letter to John Stewart (1754), in J. Y. T. Greig, *The Letters of David Hume*, vol. 1 (New York: Oxford University Press, 2011).

8. See Lawrence M. Krauss, *A Universe from Nothing: Why There Is Something Rather Than Nothing* (New York: Free Press, 2012).

9. Michael G. Strauss, "A Universe from Nothing?," *Dr. Michael G. Strauss* (blog), March 15,

2017, michaelgstrauss.com/2017/03/a-universe-from-nothing.html.

10. Clara Moskowitz, "What Is Nothing? Physicists Debate," *Live Science*, March 22, 2013, livescience.com/28132-what-is-nothing-physicists-debate.html.

11. Paul Copan, "If God Made the Universe, Who Made God?," *Enrichment* (Spring 2012): enrichmentjournal.ag.org/Issues/2012/Spring-2012/If-God-Made-the-Universe-Who-Made-God. Copan's helpful article influenced my thinking on this subject.

12. See Stephen Hawking and Leonard Mlodinow, *The Grand Design* (New York: Bantam, 2010).

13. Strauss, "A Universe from Nothing?"

14. William Lane Craig and Quentin Smith, *Theism, Atheism and Big Bang Cosmology* (Oxford: Oxford University Press, 1993), 135.

15. Michael Martin, ed., *The Cambridge Companion to Atheism* (New York: Cambridge University Press, 2007), chap. 11.

16. Derek Parfit, "Why Anything? Why This?," *London Review of Books* 20, no. 2 (January 22, 1998): lrb.co.uk/the-paper/v20/n02/derek-parfit/why-anything-why-this.

17. J. Oliver Buswell, "The Place of Paradox in Our Christian Testimony," *Journal of the American Scientific Affiliation* 17 (September 1965): 88–92, 96, asa3.org/ASA/PSCF/1965/JASA9-65complete.pdf.

18. Sean M. Carroll, "Why Is There Something, Rather Than Nothing?," (February 8, 2018): doi:10.48550/arXiv.1802.02231.

19. Nick Rose, "What Is Nothing?," *Vice*, posted October 31, 2018, vice.com/en_us/article/vbk5va/what-is-nothing.

20. Lee Strobel, "The Evidence of Cosmology: Beginning with a Bang," an interview with William Lane Craig, in *The Case for the Creator* (Grand Rapids, MI: Zondervan, 2004), 109.

21. Gottfried Wilhelm Leibniz, "The Principles of Nature and of Grace, Based on Reason," in *Philosophic Classics: Bacon to Kant*, ed. Walter Kaufmann (Englewood Cliffs, NJ: Prentice Hall, 1961), 7:256.

22. See Kenneth Richard Samples, *Christianity Cross-Examined: Is It Rational, Relevant, and Good?* (Covina, CA: RTB Press, 2021), chap. 1.

23. See Hugh Ross, *The Creator and the Cosmos: How the Latest Scientific Discoveries Reveal God*, 4th ed. (Covina, CA: RTB Press, 2018).

24. Stephen W. Hawking and Roger Penrose, *The Nature of Space and Time* (Princeton, NJ: Princeton University Press, 1996), 20.

25. John D. Barrow and Joseph Silk, *The Left Hand of Creation: The Origin and Evolution of the Expanding Universe* (Oxford: Oxford University Press, 1994), 38.

26. Anthony Kenny, *The Five Ways: St. Thomas Aquinas' Proofs of God's Existence* (Oxford: Oxford University Press, 1994), 66.

27. Strauss, "A Universe from Nothing?"

28. Michael G. Strauss, "Was the Big Bang Really the Beginning?," *Dr. Michael G. Strauss* (blog), April 27, 2019, michaelgstrauss.com/2019/04/was-big-bang-really-beginning.html.

29. Strauss, "A Universe from Nothing?"

30. Jeffrey A. Zweerink, *Who's Afraid of the Multiverse?* (Covina, CA: Reasons to Believe, 2008).

31. See Alex Vilenkin, *Many Worlds in One: The Search for Other Universes* (New York: Hill and Wang, 2006).

32. I have written on the multiverse and its challenges in Kenneth Richard Samples, *7 Truths That Changed the World: Discovering Christianity's Most Dangerous Ideas* (Grand Rapids, MI: Baker Books, 2012), 111–112.

33. George Ellis, "Does the Multiverse Really Exist?," *Scientific American,* August 2011, scientificamerican.com/article/does-the-multiverse-really-exist/.

34. See Kenneth Richard Samples, *Without a Doubt: Answering the 20 Toughest Faith Questions* (Grand Rapids, MI: Baker Books, 2004), chap. 6.

35. Vilenkin, *Many Worlds in One,* 176.

36. Alex Vilenkin, cited in Lisa Grossman, "Why Physicists Can't Avoid a Creation Event," *New Scientist,* January 11, 2012, newscientist.com/article/mg21328474-400-why-physicists-cant-avoid-a-creation-event/.

37. Jeff Zweerink, "Is There Life Out There?," interview by Sean McDowell (blog), posted October 26, 2017, seanmcdowell.org/blog/is-there-life-out-there-interview-with-astrophysicist-jeff-zweerink.

38. Paul Davies, "A Brief History of the Multiverse," *New York Times,* April 12, 2003, nytimes.com/2003/04/12/opinion/a-brief-history-of-the-multiverse.html.

39. Richard Swinburne and Steve L. Porter, "Swinburne: On Arguments for God's Existence," *The Table,* September 13, 2012, cct.biola.edu/swinburne-on-arguments-for-god-s-existence/.

40. As quoted in John J. Pasquini, *The Existence of God: Convincing and Converging Arguments* (Lanham, ND: University Press of America, 2010), 18.

41. Malcolm M. Browne, "Clues to the Universe's Origin Expected," *New York Times,* March 12, 1978, nytimes.com/1978/03/12/archives/clues-to-universe-origin-expected-the-making-of-the-universe.html.

42. For further discussion of how big bang cosmology corresponds to the biblical doctrine of creation *ex nihilo,* see Samples, *7 Truths That Changed the World,* chaps. 5–6.

43. J. I. Packer, *Concise Theology: A Guide to Historic Christian Beliefs* (Wheaton, IL: Tyndale, 1993), 26.

44. Copan, "If God Made the Universe, Who Made God?"

Chapter 3 – Creation of the Cosmos
by Hugh Ross

1. One example is Bruce K. Waltke, *Creation and Chaos: An Exegetical and Theological Study of Biblical Cosmogony* (Portland, OR: Western Conservative Baptist Seminary, 1974). Waltke has since prepared a 900-page manuscript on the first three verses of Genesis 1.

2. Hugh Ross, *The Fingerprint of God,* commemorative ed. (Glendora, CA: Reasons to Believe, 2010), 21–27.

3. Hermann Bondi and T. Gold, "The Steady-State Theory of the Expanding Universe," *Monthly Notices of the Royal Astronomical Society* 108 (1948): 252–270, doi:10.1093/mnras/108.3.252; Fred Hoyle, "A New Model for the Expanding Universe," *Monthly Notices of the Royal Astronomical Society* 108 (1948): 372–382, doi:10.1093/mnras/108.5.372; Ross, *Fingerprint of God,* 53–105.

4. Robert H. Dicke et al., "Cosmic Black-Body Radiation," *Astrophysical Journal* 142 (1965): 414–419, doi:10.1086/148306; John Gribbin, "Oscillating Universe Bounces Back," *Nature* 259 (1976): 15–16.

5. Genesis 1:1; Psalm 33:6–9; 90:2; John 17:24; 2 Timothy 1:9; Titus 1:2; Revelation 21:1.

6. Job 37:23; Jeremiah 23:24; John 1:3; Ephesians 1:4; Colossians 1:15–16; 2 Timothy 1:9; Titus 1:2; Hebrews 11:3.

7. Thomas E. McComiskey in R. Laird Harris, Gleason L. Archer Jr., and Bruce K. Waltke, *Theological Wordbook of the Old Testament,* vol. 1 (Chicago: Moody Press, 1980), 127.

8. McComiskey, *Theological Wordbook*, 127.

9. McComiskey, *Theological Wordbook*, 127.

10. Genesis 2:3–4; Psalm 33:6; 102:25; 148:5; Isaiah 40:26; 42:5; 45:18; John 1:3; Colossians 1:15–17; 2 Timothy 1:9; Titus 1:2.

11. R. Laird Harris, Gleason L. Archer Jr., and Bruce K. Waltke, *Theological Wordbook of the Old Testament*, vol. 2 (Chicago: Moody Press, 1980), 701–702.

12. Harris, Archer, and Waltke, *Theological Wordbook*, vol. 1, 213–214.

13. Harris, Archer, and Waltke, *Theological Wordbook*, vol. 1, 199.

14. Harris, Archer, and Waltke, *Theological Wordbook*, vol. 2, 608.

15. Harris, Archer, and Waltke, *Theological Wordbook*, vol. 2, 823.

16. Harris, Archer, and Waltke, *Theological Wordbook*, vol. 1, 393.

17. Harris, Archer, and Waltke, *Theological Wordbook*, vol. 1. There were no vowels in the oldest Hebrew biblical manuscripts. The later insertion of vowel sounds multiplied the 3,067 biblical Hebrew words (not including seldom used proper nouns) to 8,674 (including all proper nouns used in the Old Testament).

18. Harris, Archer, and Waltke, *Theological Wordbook*, vol. 1, 74–75.

19. Harris, Archer, and Waltke, *Theological Wordbook*, vol. 2, 935–936.

20. 2 Corinthians 12:2.

21. Waltke, *Creation and Chaos*, 20, 25–26. This point was also one of Waltke's central themes in his Kenneth S. Kantzer Lectures in Systematic Theology given January 8–10, 1991, at Trinity Evangelical Divinity School in Deerfield, Illinois; Allen P. Ross, *Creation and Blessing: A Guide to the Study and Exposition of Genesis* (Grand Rapids, MI: Baker Books, 1988), 721, 725–726.

22. Genesis 1:1; Psalm 33:6–9; Isaiah 40:26–28; 42:5; John 1:3; 17:24; Ephesians 1:4; Colossians 1:15–16; 2 Timothy 1:9; Titus 1:2; Hebrews 11:3; 1 Peter 1:20; Revelation 4:11.

23. Stephen Hawking and Roger Penrose, "The Singularities of Gravitational Collapse and Cosmology," *Proceedings of the Royal Society of London A* 314 (January 27, 1970): 529–548, doi:10.1098/rspa.1970.0021.

24. Arvind Borde, Alan H. Guth, and Alexander Vilenkin, "Inflationary Spacetimes Are Incomplete in Past Directions," *Physical Review Letters* 90, no. 15 (April 18, 2003): 151301, doi:10.1103/PhysRevLett.90.151301.

25. Psalm 119:160; Ecclesiastes 7:1–25; Acts 17:11; Romans 12:2; 1 Thessalonians 5:21; Hebrews 6:18; 1 John 4:1.

26. Isaiah 41:5–7; 44:9–20; Jeremiah 23:9–40; Colossians 2:4, 8.

27. David Toshio Tsumura, *The Earth and the Waters in Genesis 1 and 2: A Linguistic Investigation* (Sheffield, UK: Sheffield Academic, 1989), 41–43.

28. C. John Collins, *Genesis 1–4: A Linguistic, Literary, and Theological Commentary* (Phillipsburg, NJ: P&R, 2006), 51.

29. Collins, *Genesis 1–4*, 51.

30. Rodney Whitefield, *Genesis One and the Age of the Earth: What Does the Bible Say?* (San Jose, CA: R. Whitefield, 2011), 10–11. The entire booklet is available as a free download at creationingenesis.com/booklet. html.

31. Whitefield, *Genesis One*, 10–17.

32. Herbert W. Morris, *Work-Days of God: or Science and the Bible*, enlarged ed. (London: W. Nicholson and Sons, 1915), 21–106; John Cunningham Geikie, *Hours with the Bible*, vol. 1 (New York: James Pott, 1905), 40–42.

33. C. I. Scofield, *The Scofield Reference Bible* (New York: Oxford University Press, 1945), 3–4.

34. James Buswell, Hugh Ross, Robert Saucy, and Dallas Willard, *Round Table on Genesis One*, 120-minute videocassette (Pasadena, CA: Reasons to Believe, 1992). Four scholars, including gap theorist Robert Saucy, interact on their differing interpretations of Genesis 1.

35. Bernard Ramm, *The Christian View of Science and Scripture* (Grand Rapids, MI: Eerdmans, 1954), 195–210.

36. Barbara Ercolano, Antonia Bevan, and Thomas Robitaille, "The Spectral Energy Distribution of Protoplanetary Discs around Massive Young Stellar Objects," *Monthly Notices of the Royal Astronomical Society* 428 (January 2013): 2714–2722, doi:10.1093/mnras/sts249; S. T. Megeath et al., "The Spitzer Space Telescope Survey of the Orion A and B Molecular Clouds—Part I: A Census of Dusty Young Stellar Objects and a Study of Their Mid-Infrared Variability," *Astronomical Journal* 144 (December 2012): id. 192, doi:10.1088/0004-6256/144/6/192; Matthew S. Povich et al., "A Pan-Carina Young Stellar Object Catalog: Intermediate-Mass Young Stellar Objects in the Carina Nebula Identified via Mid-Infrared Excess Emission," *Astrophysical Journal Supplement* 194 (May 2011): id. 14, doi:10.1088/0067-0049/194/1/14; Woojin Kwon, "Circumstellar Structure Properties of Young Stellar Objects: Envelopes, Bipolar Outflows, and Disks" (PhD dissertation, University of Illinois at Urbana-Champaign, December 2009), ideals.illinois.edu/bitstream/handle/2142/14711/Kwon_Woojin.pdf?sequence=2; J. M. De Buizer, "Testing the Circumstellar Disc Hypothesis: A Search for H_2 Outflow Signatures from Massive Young Stellar Objects with Linearly Distributed Methanol Masers," *Monthly Notices of the Royal Astronomical Society* 341, no. 1 (May 2003): 277–298, doi:10.1046/j.1365-8711.2003.06419.x; C. Bertout, "Occultation of Young Stellar Objects by Circumstellar Disks. I. Theoretical Expectations and Preliminary Comparison with Observations," *Astronomy and Astrophysics* 363 (November 2000): 984–990, bibcode:2000A&A...363..984B.

37. C. P. Dullemond and J. D. Monnier, "The Inner Regions of Protoplanetary Disks," *Annual Review of Astronomy and Astrophysics* 48 (2010): 205–239, doi:10.1146/annurev-astro-081309-130932; Mark C. Wyatt, "Evolution of Debris Disks," *Annual Review of Astronomy and Astrophysics* 46 (2008): 339–383, doi:10.1146/annurev.astro.45.051806.110525.

38. Exoplanet TEAM, Catalog, *The Extrasolar Planets Encyclopaedia* (May 24, 2022), http://exoplanet.eu/catalog.

39. Eliza Miller-Ricci and Jonathan J. Fortney, "The Nature of the Atmosphere of the Transiting Super-Earth GJ 1214b," *Astrophysical Journal Letters* 716 (June 10, 2010): L74–L79, doi:10.1088/2041-8205/716/1/L74.

40. P. Jonathan Patchett, "Scum of the Earth After All," *Nature* 382 (August 29, 1996): 758.

41. Harris, Archer, and Waltke, *Theological Wordbook*, vols. 1 and 2. These volumes contain definitions for 3,067 Hebrew words including Old Testament names that have theological import such as Abraham, David, Jerusalem, Jordan, and Sinai.

42. Harris, Archer, and Waltke, *Theological Wordbook*, vol. 1, 370–371; William Wilson, *Old Testament Word Studies* (Grand Rapids, MI: Kregel, 1978), 109.

43. Harris, Archer, and Waltke, *Theological Wordbook*, vol. 1, 370–371, vol. 2, 672–673; H. W. F. Gesenius, *Gesenius' Hebrew-Chaldee Lexicon to the Old Testament* (Grand Rapids, MI: Baker Books, 1979), 612–613.

44. R. Monastersky, "Speedy Spin Kept Early Earth from Freezing," *Science News* 143 (June 12, 1993): 373, doi:10.2307/3977267; William R. Kuhn, J. C. G. Walker, and Hal G. Marshall, "The Effect on Earth's Surface Temperature from Variations in Rotation Rate, Continent Formation, Solar Luminosity, and Carbon Dioxide," *Journal of Geophysical Research: Atmospheres* 94 (August 20, 1989): 11, 129–131, 136, doi:10.1029/jd094id08p11129; George E. Williams,

"Geological Constraints on the Precambrian History of Earth's Rotation and the Moon's Orbit," *Reviews of Geophysics* 38 (February 2000): 37–60, doi:10.1029/1999RG900016.

45. Kuhn, Walker, and Marshall, "The Effect on Earth's Surface Temperature," 11, 129–131, 136; Hubert P. Yockey, *Information Theory and Molecular Biology* (Cambridge, UK: Cambridge University Press, 1992), 222–223; Richard A. Kerr, "Fiery Io Models Earth's First Days," *Science* 280, no. 5362 (April 17, 1998): 382, doi:10.1126/science.280.5362.381.

46. Yuichiro Ueno et al., "Ion Microprobe Analysis of Graphite from Ca. 3.8 Ga Metasediments, Isua Supracrustal Belt, West Greenland: Relationship between Metamorphism and Carbon Isotopic Composition," *Geochimica et Cosmochimica Acta* 66 (2002): 1257–1268: doi:10.1016/S0016-7037(01)00840-7; Minik T. Rosing, "^{13}C-Depleted Carbon Microparticles in >3700-Ma Sea-Floor Sedimentary Rocks from West Greenland," *Science* 283 (January 29, 1999): 674–676, doi:10.1126/science.283.5402.674; Craig E. Manning et al., "Geology and Age of Supracrustal Rocks: Akilia Island, Greenland: New Evidence for a >3.83 Ga Origin of Life," *Astrobiology* 1 (2001): 402–403; C. M. Fedo, J. S. Meyers, and P. W. U. Appel, "Depositional Setting and Paleogeographic Implications of Earth's Oldest Supracrustal Rocks, the >3.7 Ga Isua Greenstone Belt, West Greenland," *Sedimentary Geology* 141–142 (2001): 61–77, doi:10.1016/S0037-0738(01)00068-9; J. O'Neil and R. W. Carlson, "A Glimpse of Earth's Primordial Crust: The Nuvvuagittuq Greenstone Belt as a Vestige of Mafic Hadean Oceanic Crust," *American Geophysical Union, Fall Meeting 2010* (December 2010): abstract #V43D-04; Uffe Gråe Jørgensen et al., "The Earth-Moon System during the Late Heavy Bombardment Period," *Icarus* 204 (2009): 368–380, doi:10.1016/j.icarus.2009.07.015.

Chapter 4 – Why Such a Vast Universe?
by Hugh Ross

1. See Hugh Ross, *Why the Universe Is the Way It Is* (Grand Rapids, MI: Baker Books, 2008), 20.

2. Victor J. Stenger, *God: The Failed Hypothesis: How Science Shows That God Does Not Exist* (Amherst, NY: Prometheus, 2007), 156.

3. Stephen W. Hawking, *A Brief History of Time* (New York: Bantam Books, 1998), 126.

4. Stenger, *God*, 157.

5. Stenger, *God*, 157.

6. For more on this topic of optimization, see Ross, *Why the Universe Is the Way It Is.*

7. See Ross, *Why the Universe Is the Way It Is*, chaps. 3 and 5.

8. Steven V. W. Beckwith et al., "The Hubble Ultra Deep Field," *Astronomical Journal* 132 (November 2006): 1729–1755.

9. See Ross, *The Creator and the Cosmos: How the Latest Scientific Discoveries Reveal God,* 4th ed. (Covina, CA: RTB Press, 2018), 124.

10. See Masataka Fukugita and P. J. E. Peebles, "The Cosmic Energy Inventory," *Astrophysical Journal* 616 (December 2004): 643–668, doi:10.1086/425155; D. N. Spergel et al., "Three-Year *Wilkinson Microwave Anisotropy Probe (WMAP)* Observations: Implications for Cosmology," *Astrophysical Journal Supplement Series* 170 (June 2007): 377–408, doi:10.1086/513700.

11. See Ross, *Why the Universe Is the Way It Is*, appendix B.

12. See Ross, *Why the Universe Is the Way It Is*, appendix B.

13. See Ross, *Why the Universe Is the Way It Is*, appendix B.

14. See Peter Coles and George F. R. Ellis, *Is the Universe Open or Closed? The Density of Matter in the Universe* (Cambridge: Cambridge University Press, 1997); Peter Coles, ed., *The*

Routledge Critical Dictionary of the New Cosmology (New York: Routledge, 1998), 180–183; Lawrence M. Krauss, "The End of the Age Problem and the Case for a Cosmological Constant Revisited," *Astrophysical Journal* 501 (July 10, 1998): 461, 465.

15.   Fukugita and Peebles, "Cosmic Energy Inventory," 643–668.

16.   Spergel et al., "Three-Year *Wilkinson Microwave Anisotropy Probe (WMAP)* Observations," 377–408; E. Komatsu et al., "Five-Year Wilkinson Microwave Anisotropy Probe (WMAP) Observations: Cosmological Interpretation" (preprint, National Aeronautics and Space Administration, 2008), lambda.gsfc.nasa.gov/product/map/dr3/pub_papers/fiveyear/cosmology/wmap_5yr_cosmo.pdf.

17.   John North, *The Norton History of Astronomy and Cosmology* (New York: W. W. Norton, 1994), 502–507; P. J. E. Peebles, *Principles of Physical Cosmology* (Princeton, NJ: Princeton University Press, 1993), 417.

18.   Morton S. Roberts, "The Content of Galaxies: Stars and Gas," in *Annual Review of Astronomy and Astrophysics*, ed. Leo Goldberg, Armin J. Deutsch, and David Layzer (Palo Alto, CA: Annual Reviews, 1963), 160–163; J. H. Oort, "Stellar Dynamics," in *Galactic Structure*, ed. Adriaan Blaauw and Maarten Schmidt (Chicago: University of Chicago Press, 1965), 469–473; Rudolf Kippenhahn and Alfred Weigert, *Stellar Structure and Evolution*, corrected printing (New York: Springer-Verlag, 1994), 268.

19.   Additional fine-tuning requirements for the quantities and locations of the different forms of matter are addressed in Ross, *Why the Universe Is the Way It Is*, chap. 3.

20.   Kippenhahn and Weigert, *Stellar Structure and Evolution*, 215, 266–269.

21.   Ruth A. Daly and S. G. Djorgovski, "Direct Determination of the Kinematics of the Universe and Properties of the Dark Energy as Functions of Redshift," *Astrophysical Journal* 612 (September 10, 2004): 652–659, doi:10.1086/422673; Ruth A. Daly et al., "Improved Constraints on the Acceleration History of the Universe and the Properties of the Dark Energy," *Astrophysical Journal* 677 (April 10, 2008): 1–11, doi:10.1086/528837.

22.   The natural consequences for life in a universe that can die (from a "heat death" caused by continuing expansion making the universe colder and colder) but can never be "reborn" are addressed in Ross, *Why the Universe Is the Way It Is*, chap. 6.

23.   For a more thorough analysis of the demise of the oscillating universe model and of the Hindu / Buddhist / New Age concept of a reincarnating universe, see Ross, *Creator and the Cosmos*, 48–67, 87–98, 169–174.

24.   See Ross, *Why the Universe Is the Way It Is*, appendix B.

25.   Bradley E. Schaefer, "The Hubble Diagram to Redshift >6 from 69 Gamma-Ray Bursts," *Astrophysical Journal* 660 (May 2007): 16–46, doi:10.1086/511742; Uros Seljak, Anze Slosar, and Patrick McDonald, "Cosmological Parameters from Combining the Lyman-α Forest with CMB, Galaxy Clustering and SN Constraints," *Journal of Cosmology and Astroparticle Physics* 10 (October 19, 2006): 014, doi:10.1088/1475-7516/2006/10/014.

26.   Lawrence M. Krauss, *Quintessence: The Mystery of the Missing Mass in the Universe* (New York: Basic Books, 2000), 103–105; Krauss, "End of the Age Problem," 461, 465.

Chapter 5 – The Origin and Design of the Universe
by Jeff Zweerink

1.   See "'A Universe from Nothing' by Lawrence Krauss, AAI 2009," YouTube video, 1:04:51, posted by Richard Dawkins Foundation for Reason and Science, October 21, 2009, youtube.com/watch?v=7ImvlS8PLIo#t=16m49s.

2.   John D. Barrow and Frank J. Tipler, *The Anthropic Cosmological Principle* (Oxford: Oxford

University Press, 1986), 253.

3.   Francis Crick, *What Mad Pursuit: A Personal View of Scientific Discovery* (New York: Basic Books, 1988), 138.

4.   C. S. Lewis, *God in the Dock* (Grand Rapids, MI: Eerdmans, 1970), 53.

Chapter 6 – Would the Discovery of ET Disprove Christianity and Must ET Exist?
by Jeff Zweerink

1.   Praying Together, English Language Liturgical Consultation (1988), englishtexts.org/the-apostles-creed.

2.   See Fazale Rana and Hugh Ross, *Origins of Life: Biblical and Evolutionary Models Face Off* (Covina, CA: RTB Press, 2014); and Hugh Ross, *Improbable Planet: How Earth Became Humanity's Home* (Grand Rapids, MI: Baker Books, 2016).

3.   For a starting point into the theological significance of finding intelligent ET, see Ted Peters, "The Implications of the Discovery of Extra-Terrestrial Life for Religion," *Philosophical Transactions of the Royal Society A* 369 (February 2011): 644–655, doi:10.1098/rsta.2010.0234.

4.   For a discussion of the ways Christianity can incorporate intelligent ET, see Jeff Zweerink, *Is There Life Out There?* (Covina, CA: RTB Press, 2017), 162.

5.   For one quick example, see "Interpretations of Quantum Mechanics," *Wikipedia*, last modified April 25, 2017, en.wikipedia.org/wiki/Interpretations_of_quantum_mechanics. I realize *Wikipedia* may have some scholarly deficiencies, but this page lists at least 13 different interpretations of quantum mechanics and provides links to investigate those interpretations more thoroughly.

6.   David Sobral et al., "A Large Hα Survey at z = 2.23, 1.47, 0.84 and 0.40: The 11 Gyr Evolution of Star-Forming Galaxies from HiZELS," *Monthly Notices of the Royal Astronomical Society* 428 (January 2013): 1128–1146, doi:10.1093/mnras/sts096.

7.   *Contact*, directed by Robert Zemeckis (Burbank, CA: Warner Home Video, 1997), DVD.

8.   Rana and Ross, *Origins of Life*, 75–79.

9.   J. Alexis P. Rodriguez et al., "Tsunami Waves Extensively Resurfaced the Shorelines of an Early Martian Ocean," *Scientific Reports* 6 (2016): id. 25106, doi:10.1038/srep25106.

10.   W. T. Pike et al., "Quantification of the Dry History of the Martian Soil Inferred from In Situ Microscopy," *Geophysical Research Letters* 38 (December 2011): id. L24201, doi:10.1029/2011GL049896.

11.   Amos Banin, "The Enigma of the Martian Soil," *Science* 309 (August 2005): 888–890, doi:10.1126/science.1112794; Donald M. Hassler et al., "Mars' Surface Radiation Environment Measured with the Mars Science Laboratory's Curiosity Rover," *Science* 343 (January 2014): id. 1244797, doi:10.1126/science.1244797.

12.   *Contact*, 1997.

Chapter 7– Why I Believe God Exists: A Biochemical Case for the Creator
by Fazale Rana

1.   Richard Dawkins, *The Blind Watchmaker: Why the Evidence of Evolution Reveals a Universe without Design* (New York: W. W. Norton, 1987), 10.

2.   Cary Funk and Becka A. Alper, "Highly Religious Americans Are Less Likely Than Others to See Conflict between Faith and Science," Pew Research Center (October 22, 2015), pewinternet.org/2015/10/22/science-and-religion/.

3.   Will Clifford, "DNA Computing: Meet Dr. Adleman," Youngzine, February 2, 2013,

youngzine.org/news/technology/dna-computing-meet-dr-adleman.

4. Dawkins, *The Blind Watchmaker*, 7.
5. See Anjeanette Roberts, "Unequivocating Evolution," in *Building Bridges: A Presentation of RTB's Testable Creation Model* (Covina, CA: RTB Press, 2018).
6. Leslie Orgel, "The RNA World and the Origin of Life" (abstract, ISSOL 2002 meeting, Oaxaca, MX).
7. Leslie Orgel, "Self-Organizing Biochemical Cycles," *Proceedings of the National Academy of Sciences, USA* 97 (November 7, 2000): 12503–12507, doi:10.1073/pnas.220406697.
8. Paul Davies, *The Fifth Miracle: The Search for the Origin and Meaning of Life* (New York: Touchstone, 1999), 131.
9. Simon Conway Morris, *Life's Solution: Inevitable Humans in a Lonely Universe* (New York: Cambridge University Press, 2003), 41.
10. Lin Jiang et al., "De Novo Computational Design of Retro-Aldol Enzymes," *Science* 319 (March 7, 2008): 1387–1391, doi:10.1126/science.1152692.

Chapter 8 – The Case for the Image of God
by Fazale Rana with Kenneth Richard Samples

1. Charles Darwin, *The Descent of Man, and Selection in Relation to Sex*, 2nd ed., with an introduction by H. James Birx, Great Minds Series (Amherst, NY: Prometheus Books, 1998), 188.
2. See Fazale R. Rana with Kenneth R. Samples, *Humans 2.0: Scientific, Philosophical, and Theological Perspectives on Transhumanism* (Covina, CA: RTB Press, 2019), chap. 6.
3. See Rana with Samples, *Humans 2.0*, chap. 7.
4. C. John Collins, *Science and Faith: Friends or Foes?* (Wheaton, IL: Crossway, 2003), 124–127.
5. Kenneth R. Samples, *A World of Difference: Putting Christian Truth-Claims to the Worldview Test* (Grand Rapids, MI: Baker Books, 2007), 171–188; Kenneth R. Samples, *7 Truths that Changed the World: Discovering Christianity's Most Dangerous Ideas* (Grand Rapids, MI: Baker Books, 2012), 163–190.
6. Millard J. Erickson, *Christian Theology*, 2nd ed. (Grand Rapids, MI: Baker Books, 1998), 517–536; Wayne Grudem, *Systematic Theology: An Introduction to Biblical Doctrine* (Grand Rapids, MI: Zondervan, 1994), 442–450.
7. Thomas Suddendorf, *The Gap: The Science of What Separates Us from Other Animals* (New York: Basic Books, 2013), 217–218.
8. Alejandro Pérez, Manuel Carreiras, and Jon Andoni Duñabeitia, "Brain-to-Brain Entrainment: EEG Interbrain Synchronization While Speaking and Listening," *Scientific Reports* 7 (June 23, 2017): id. 4190, doi:10.1038/s41598-017-04464-4.
9. Suddendorf, *The Gap*, 216.
10. Fazale Rana with Hugh Ross, *Who Was Adam? A Creation Model Approach to the Origin of Humanity*, exp. and updated ed. (Covina, CA: RTB Press, 2015), 319.
11. Rana with Ross, *Who Was Adam?*, 195–201, 313–325. See also these articles by Fazale Rana: "Did Neanderthals Make Art?," *Today's New Reason to Believe* (blog), January 22, 2015, reasons.org/explore/blogs/todays-new-reason-to-believe/read/tnrtb/2015/01/22/did-neanderthals-make-art ; "Did Neanderthals Make Jewelry?," *The Cell's Design* (blog), September 28, 2016, reasons.org/explore/blogs/the-cells-design/read/the-cells-design/2016/09/28/did-neanderthals-make-jewelry; "Did Neanderthals Bury Their Dead with Flowers?," *Today's New Reason to Believe* (blog), May 9, 2016, reasons.org/explore/blogs/todays-new-reason-to-believe/read/tnrtb/2016/05/09/did-neanderthals-bury-their-dead-with-flowers; "Do Neanderthal Cave Structures Challenge Human Exceptionalism?," *Today's New Reason*

to Believe (blog), June 16, 2016, reasons.org/explore/blogs/todays-new-reason-to-believe/read/tnrtb/2016/06/16/do-neanderthal-cave-structures-challenge-human-exceptionalism; "One-of-a-Kind: Three Discoveries Affirm Human Uniqueness," *Today's New Reason to Believe* (blog), June 8, 2015, reasons.org/explore/blogs/todays-new-reason-to-believe/read/tnrtb/2015/06/08/one-of-a-kind-three-discoveries-affirm-human-uniqueness.

12. Jon Mooallem, "Neanderthals Were People, Too," *New York Times Magazine*, January 11, 2017, nytimes.com/2017/01/11/magazine/neanderthals-were-people-too.html.

13. Mooallem, "Neanderthals Were People, Too."

14. Mooallem, "Neanderthals Were People, Too."

15. Rana with Ross, *Who Was Adam?*, 200–201.

16. Rana with Ross, *Who Was Adam?*, 293–294.

17. Rana with Ross, *Who Was Adam?*, 94–295.

18. Rana with Ross, *Who Was Adam?*, 296–299.

19. Massachusetts Institute of Technology, "The Rapid Rise of Human Language," ScienceDaily (website), posted March 31, 2015, sciencedaily.com/releases/2015/03/150331131324.htm.

20. Johan J. Bolhuis et al., "How Could Language Have Evolved?," *PLoS Biology* 12 (August 26, 2014): e1001934, doi:10.1371/journal.pbio.1001934.

21. Suddendorf, *The Gap*, 2.

22. Bolhuis et al., "How Could Language Have Evolved?," e1001934.

23. Wesley J. Smith, "More Than in 'God's Image,'" *First Things* (website), July 24, 2015, firstthings.com/web-exclusives/2015/07/more-than-in-gods-image.

Index

About the Authors

Fazale "Fuz" R. Rana is president and CEO of Reasons to Believe (RTB) and holds a PhD in chemistry with an emphasis in biochemistry from Ohio University.

Fuz conducted postdoctoral work at the Universities of Virginia and Georgia and worked for seven years as a senior scientist in product development for Procter & Gamble. He has published articles in peer-reviewed scientific journals, delivered presentations at international scientific meetings, and addressed the relationship between science and Christianity at churches and universities in the US and abroad. Since joining RTB in 1999, Fuz has participated in numerous podcasts and videos and authored countless blog articles and several books, including *Humans 2.0*, *Creating Life in the Lab*, and *Fit for a Purpose*.

Fuz and his wife, Amy, live in Southern California.

Hugh Ross is senior scholar and founder of Reasons to Believe (RTB), an organization dedicated to demonstrating the compatibility of science and the Christian faith.

With a degree in physics from the University of British Columbia and a PhD in astronomy from the University of Toronto, Hugh continued his research on quasars and galaxies as a postdoctoral fellow at the California Institute of Technology before transitioning to full-time ministry. In addition to founding and leading RTB, he remains on the pastoral staff at Christ Church Sierra Madre. His writings include journal and magazine articles and numerous books—*The Creator and the Cosmos, Why the Universe Is the Way It Is*, and *Improbable Planet*, among others. He has spoken on hundreds of university campuses as well as at conferences and churches around the world.

Hugh lives in Southern California with his wife, Kathy.

Kenneth Richard Samples serves as senior research scholar at Reasons to Believe (RTB), an organization that researches and communicates how God's revelation in the Bible harmonizes with science and philosophy.

Kenneth holds a BA in social science with an emphasis in history and philosophy from Concordia University and an MA in theological studies from Talbot School of Theology. Prior to joining RTB, he worked as senior research consultant and correspondence editor at the Christian Research Institute and regularly cohosted popular call-in radio program *The Bible Answer Man*.

Kenneth's other books include *Christianity Cross-Examined*, *Classic Christian Thinkers*, and *God among Sages*. He leads RTB's *Straight Thinking* podcast and writes *Reflections*, a weekly blog dedicated to exploring the Christian worldview. Additionally, he is currently an adjunct instructor of apologetics at Biola University and has spoken at universities and churches around the world.

Kenneth lives in Southern California with his wife, Joan. They have three grown children.

Astrophysicist **Jeff Zweerink** is a senior research scholar at Reasons to Believe (RTB), an organization dedicated to demonstrating the compatibility of science and the Christian faith. He earned a PhD in astrophysics with a focus on gamma rays from Iowa State University. Jeff was involved in research projects such as STACEE and VERITAS and is coauthor of more than 30 academic papers. He is also the author of *Escaping the Beginning?*, *Is There Life Out There?*, and *Who's Afraid of the Multiverse?* Today, Jeff writes and speaks on the compatibility of science and the Christian faith, and works on GAPS (a balloon experiment seeking to detect dark matter).

Jeff and his wife, Lisa, live in Southern California and have five children.

About Reasons to Believe

RTB is an international, interdenominational ministry, whose purpose is to show that science and the Christian faith are allies, not enemies. Our passion is to bring the Christian gospel to a dying world and to inspire Christians with compelling and engaging scientific truths.

Our scholar team, consisting of three PhD scientists, a philosopher-theologian, and a philosopher-ethicist, offers distinctive and fascinating insights on topics ranging from biblical creation to cutting-edge biotechnology that bolster confidence in the veracity of the Bible and faith in the personal, transcendent God revealed in both Scripture and nature.

For more information, visit reasons.org.

For inquiries, contact us via:
818 S. Oak Park Rd. Covina, CA 91724
(855) REASONS | (855) 732-7667
ministrycare@reasons.org

IF GOD MADE THE UNIVERSE . . .

HUGH ROSS

WHY THE UNIVERSE IS THE WAY IT IS

Why is it so vast? Why allow decay and death to be part of it? Why wait to bring humans onto the scene? Why let one man and one woman ruin it for everyone? Why isn't the meaning of my life more obvious?

Explore these questions and more in Why the Universe Is the Way It Is. Get your copy now at **support.reasons.org** or wherever you buy books.

Scientific, philosophical, and theological perspectives on transhumanism

Should we discourage advances in biotechnology and bioengineering that can be used for human enhancement? Or should we take control of our own "evolution" and usher in a posthuman age? Is there another option?

Examine the challenging topic of transhumanism in *Humans 2.0*—also available in audiobook format. Get your copy now at **support.reasons.org.**